Numerical Recipes Example Book (Pascal)

Revised Edition

Numerical Recipes Example Book (Pascal)

Revised Edition

William T. Vetterling

Polaroid Corporation

Saul A. Teukolsky

Department of Physics, Cornell University

William H. Press

Harvard-Smithsonian Center for Astrophysics

Brian P. Flannery

EXXON Research and Engineering Company

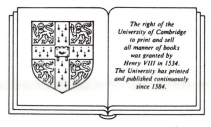

The right of the
University of Cambridge
to print and sell
all manner of books
was granted by
Henry VIII in 1534.
The University has printed
and published continuously
since 1584.

CAMBRIDGE UNIVERSITY PRESS

Cambridge

New York   Port Chester   Melbourne   Sydney

Published by the Press Syndicate of the University of Cambridge
The Pitt Building, Trumpington Street, Cambridge CB2 1RP
40 West 20th Street, New York, NY 10011 U.S.A.
10 Stamford Road, Oakleigh, Melbourne 3166, Australia

First published 1989
Reprinted 1990 (twice)

Printed in the United States of America
Typeset in T$_{\mathrm{E}}$X

The computer programs in this book, and the procedures in *Numerical Recipes in Pascal: The Art of Scientific Computing*, are available in several machine-readable formats. To purchase diskettes in IBM PC or Macintosh format, use the order form at the back of this book, or write to Cambridge University Press, 110 Midland Avenue, Port Chester, New York 10573. The main book, example books, and diskettes are available in editions in the FORTRAN, Pascal, and C languages. Technical questions, corrections, and requests for information on other available formats and additional software products should be directed to Numerical Recipes Software, P.O. Box 243, Cambridge, Massachusetts 02238. Inquiries regarding mainframe and workstation licenses should also be directed to this address.

ISBN 0 521 37675 0

CONTENTS

Preface

This *Numerical Recipes Example Book* is designed to accompany the text and reference book *Numerical Recipes in Pascal: The Art of Scientific Computing* by William H. Press, Brian P. Flannery, Saul A. Teukolsky, and William T. Vetterling (Cambridge University Press, 1989). In that volume, the algorithms and methods of scientific computation are developed in considerable detail, starting with basic mathematical analysis and working through to actual implementation in the form of Pascal procedures. The routines in *Numerical Recipes in Pascal: The Art of Scientific Computing*, numbering nearly 200, are meant to be incorporated into user applications; they are procedures (or functions), not stand-alone programs.

It often happens, when you want to incorporate somebody else's procedure into your own application program, that you first want to see the procedure demonstrated on a simple example. Prose descriptions of how to use a procedure (even those in *Numerical Recipes*) can occasionally be inexact. There is no substitute for an actual, Pascal demonstration program that shows exactly how data are fed to a procedure, how the procedure is called, and how its results are unloaded and interpreted.

Another not unusual case occurs when you have, for one seemingly good purpose or another, modified the source code in a "foreign" procedure. In such circumstances, you might well want to test the modified procedure on an example known previously to have worked correctly, *before* letting it loose on your own data. There is the related case where procedure source code may have become corrupted, e.g., lost some lines or characters in transmission from one machine to another, and a simple revalidation test is desirable.

These are the needs addressed by this *Numerical Recipes Example Book*. Divided into chapters identically with *Numerical Recipes in Pascal: The Art of Scientific Computing*, this book contains Pascal source programs that exercise and demonstrate all of the *Numerical Recipes* procedures and functions. Each program contains comments, and is prefaced by a short description of what it does, and of which *Numerical Recipes* routines it exercises. In cases where the demonstration programs require input data, that data is also printed in this book. In some cases, where the demonstration programs are not "self-validating," sample output is also shown.

Necessarily, in the interests of clarity, the *Numerical Recipes* procedures and functions are demonstrated in simple ways. A consequence is that the demonstration programs in this book do not usually test all possible regimes of input data, or even all lines of procedure source code. The demonstration programs in this book were by no means the only validating tests that the *Numerical Recipes* procedures and functions were required to pass during their development. The programs in this book *were* used during the later stages of the production of *Numerical Recipes in Pascal: The Art of*

Scientific Computing to maintain integrity of the source code, and in this role were found to be invaluable.

The `Pascal` procedures in *Numerical Recipes in Pascal: The Art of Scientific Computing* have been completely rewritten from the procedures published as an Appendix at the back of the original `FORTRAN` edition of *Numerical Recipes*. This example book is a revised version of our earlier book *Numerical Recipes Example Book (Pascal)*, and is intended to be used with the rewritten `Pascal` procedures. The example programs in this book will *not* work with the old `Pascal` procedures.

DISCLAIMER OF WARRANTY

THE PROGRAMS LISTED IN THIS BOOK ARE PROVIDED "AS IS" WITHOUT WARRANTY OF ANY KIND. WE MAKE NO WARRANTIES, EXPRESS OR IMPLIED, THAT THE PROGRAMS ARE FREE OF ERROR, OR ARE CONSISTENT WITH ANY PARTICULAR STANDARD OF MERCHANTABILITY, OR THAT THEY WILL MEET YOUR REQUIREMENTS FOR ANY PARTICULAR APPLICATION. THEY SHOULD NOT BE RELIED ON FOR SOLVING A PROBLEM WHOSE INCORRECT SOLUTION COULD RESULT IN INJURY TO A PERSON OR LOSS OF PROPERTY. IF YOU DO USE THE PROGRAMS OR PROCEDURES IN SUCH A MANNER, IT IS AT YOUR OWN RISK. THE AUTHORS AND PUBLISHER DISCLAIM ALL LIABILITY FOR DIRECT, INCIDENTAL, OR CONSEQUENTIAL DAMAGES RESULTING FROM YOUR USE OF THE PROGRAMS, SUBROUTINES, OR PROCEDURES IN THIS BOOK OR IN *Numerical Recipes: The Art of Scientific Computing*.

Chapter 1: Preliminaries

The routines in Chapter 1 of Numerical Recipes are introductory and less general in purpose than those in the remainder of the book. This chapter's routines serve primarily to expose the book's notational conventions, illustrate control structures, and perhaps to amuse. You may even find them useful. We hope that you will use badluk for no serious purpose.

⋆ ⋆ ⋆ ⋆

Procedure flmoon calculates the phases of the moon, or more exactly, the Julian day and fraction thereof on which a given phase will occur or has occurred. The program d1r1 asks the present date and compiles a list of upcoming phases. We have compared the predictions to lunar tables, with happy results. Shown are the results of a test run, which you may replicate as a check. In this program, notice that we have set zon (the time zone) to −5.0 to signify the five hour separation of the Eastern Standard time zone from Greenwich, England. (You use positive values of zon if you are east of Greenwich.) The Julian day results are converted to calendar dates through the use of caldat, which appears later in the chapter. The fractional Julian day and time zone combine to form a correction that can possibly change the calendar date by one day.

Date	Time(EST)	Phase
1 9 1982	3 PM	full moon
1 16 1982	7 PM	last quarter
1 24 1982	11 PM	new moon
2 1 1982	10 AM	first quarter
2 8 1982	2 AM	full moon
2 15 1982	3 PM	last quarter
2 23 1982	4 PM	new moon
3 2 1982	6 PM	first quarter
3 9 1982	3 PM	full moon
3 17 1982	12 AM	last quarter
3 25 1982	5 AM	new moon
4 1 1982	0 AM	first quarter
4 8 1982	5 AM	full moon
4 16 1982	7 AM	last quarter
4 23 1982	4 PM	new moon
4 30 1982	7 AM	first quarter
5 7 1982	8 PM	full moon
5 16 1982	0 AM	last quarter
5 23 1982	0 AM	new moon
5 29 1982	3 PM	first quarter

```pascal
PROGRAM d1r1(input,output);
(* driver for routine FLMOON *)
(*$I MODFILE.PAS *)
CONST
   zon = -5.0;
TYPE
   CharArray13 = PACKED ARRAY [1..13] OF char;
VAR
   timzon,frac,secs: real;
   i,i1,i2,i3,id,im,iy,n,nph: integer;
   j1,j2: longint;
   phase: ARRAY [0..3] OF CharArray13;
(*$I JULDAY.PAS *)
(*$I CALDAT.PAS *)
(*$I FLMOON.PAS *)
BEGIN
   timzon := zon/24.0;
   phase[0] := 'new moon     ';
   phase[1] := 'first quarter';
   phase[2] := 'full moon    ';
   phase[3] := 'last quarter ';
   writeln('date of the next few phases of the moon');
   writeln('enter today''s date (e.g. 1 31 1982)   :   ');
   readln(im,id,iy);
(* approximate number of full moons since january 1900 *)
   n := trunc(12.37*(iy-1900+trunc((im-0.5)/12.0)));
   nph := 2;
   j1 := julday(im,id,iy);
   flmoon(n,nph,j2,frac);
   n := n+trunc((j1-j2)/28.0);
   writeln;
   writeln('date':10,'time(est)':19,'phase':9);
   FOR i := 1 TO 20 DO BEGIN
      flmoon(n,nph,j2,frac);
      frac := 24.0*(frac+timzon);
      IF frac < 0.0 THEN BEGIN
         j2 := j2-1;
         frac := frac+24.0
      END;
      IF frac > 12.0 THEN BEGIN
         j2 := j2+1;
         frac := frac-12.0
      END ELSE
         frac := frac+12.0;
      i1 := trunc(frac);
      secs := 3600.0*(frac-i1);
      i2 := trunc(secs/60.0);
      i3 := trunc(secs-60*i2);
      caldat(j2,im,id,iy);
      writeln(im:5,id:3,iy:5,i1:9,':',i2:2,':',i3:2,' ':5,phase[nph]);
      IF nph = 3 THEN BEGIN
         nph := 0;
         n := n+1
      END ELSE
         nph := nph+1
   END
END.
```

Program julday, our exemplar of the IF control structure, converts calendar dates to Julian dates. Not many people know the Julian date of their birthday or any other convenient reference point, for that matter. To remedy this, we offer a list of checkpoints, which appears at the end of this chapter as the file dates1.dat. The program d1r2 lists the Julian date of each historic event for comparison. Then it allows you to make your own entertaining choices.

```pascal
PROGRAM d1r2(input,output,dfile);
(* driver for JULDAY *)
(*$I MODFILE.PAS *)
LABEL 99;
TYPE
   StrArray10 = string[10];
   StrArray40 = string[40];
VAR
   i,id,im,iy,n: integer;
   txt: StrArray40;
   name: ARRAY [1..12] OF StrArray10;
   dfile: text;
(*$I JULDAY.PAS *)
BEGIN
   name[1] := 'january    '; name[2] := 'february ';
   name[3] := 'march      '; name[4] := 'april       ';
   name[5] := 'may        '; name[6] := 'june        ';
   name[7] := 'july       '; name[8] := 'august      ';
   name[9] := 'september '; name[10] := 'october    ';
   name[11] := 'november  '; name[12] := 'december  ';
   NROpen(dfile,'dates1.dat');
   readln(dfile,txt);
   readln(dfile,n);
   writeln;
   writeln('month':5,'day':8,'year':6,'julian day':12,'event':9);
   writeln;
   FOR i := 1 TO n DO BEGIN
      readln(dfile,im,id,iy,txt);
      writeln(name[im]:10,id:3,iy:6,julday(im,id,iy):10,' ':5,txt);
   END;
   close(dfile);
   writeln;
   writeln('your choices:');
   writeln('month day year (e.g. 1 13 1905)');
   FOR i := 1 TO 20 DO BEGIN
      writeln;
      writeln('mm dd yyyy ?');
      readln(im,id,iy);
      IF im < 0 THEN GOTO 99;
      writeln('julian day: ',julday(im,id,iy))
   END;
99:
END.
```

The next program in *Numerical Recipes* is badluk, an infamous code that combines the best and worst instincts of man. The demonstration program for badluk is trivial, with sample results appearing in the main text.

```
PROGRAM d1r3(input,output);
(* driver for BADLUK *)
(*$I MODFILE.PAS *)
(*$I JULDAY.PAS *)
(*$I FLMOON.PAS *)
(*$I BADLUK.PAS *)
BEGIN
   badluk
END.
```

Chapter 1 closes with routine caldat, which illustrates no new points, but complements julday by doing conversions from Julian day number to the month, day, and year on which the given Julian day began. This offers an opportunity, grasped by the demonstration program d1r4, to push dates through both julday and caldat in succession, to see if they survive intact. This, of course, tests only your authors' ability to make mistakes backwards as well as forwards, but we hope you will share our optimism that correct results here speak well for both routines. (We have checked them a bit more carefully in other ways.)

```
PROGRAM d1r4(input,output,dfile);
(* driver for routine CALDAT *)
(*$I MODFILE.PAS *)
TYPE
   CharArray10 = PACKED ARRAY [1..10] OF char;
VAR
   i,id,idd,im,imm,iy,iyy,n: integer;
   j: longint;
   name: ARRAY [1..12] OF CharArray10;
   dfile: text;
(*$I JULDAY.PAS *)
(*$I CALDAT.PAS *)
BEGIN
(* check whether caldat properly undoes the operation of julday *)
   name[1]  := 'january   ';
   name[2]  := 'february  ';
   name[3]  := 'march     ';
   name[4]  := 'april     ';
   name[5]  := 'may       ';
   name[6]  := 'june      ';
   name[7]  := 'july      ';
   name[8]  := 'august    ';
   name[9]  := 'september ';
   name[10] := 'october   ';
   name[11] := 'november  ';
   name[12] := 'december  ';
   NROpen(dfile,'dates1.dat');
   readln(dfile);
   readln(dfile,n);
   writeln;
   writeln('original date:','reconstructed date:':43);
   writeln('month':5,'day':8,'year':6,'julian day':15,
      'month':9,'day':8,'year':6);
   FOR i := 1 TO n DO BEGIN
      readln(dfile,im,id,iy);
      j := julday(im,id,iy);
      caldat(j,imm,idd,iyy);
```

```
      writeln(name[im],id:3,iy:6,j:13,name[imm]:16,idd:3,iyy:6)
   END
END.
```

Appendix

File dates1.dat:

```
List of dates for testing routines in Chapter 1
16 entries
12 31   -1 End of millennium
01 01    1 One day later
10 14 1582 Day before Gregorian calendar
10 15 1582 Gregorian calendar adopted
01 17 1706 Benjamin Franklin born
04 14 1865 Abraham Lincoln shot
04 18 1906 San Francisco earthquake
05 07 1915 Sinking of the Lusitania
07 20 1923 Pancho Villa assassinated
05 23 1934 Bonnie and Clyde eliminated
07 22 1934 John Dillinger shot
04 03 1936 Bruno Hauptman electrocuted
05 06 1937 Hindenburg disaster
07 26 1956 Sinking of the Andrea Doria
06 05 1976 Teton dam collapse
05 23 1968 Julian Day 2440000
```

Chapter 2: Linear Algebraic Equations

Numerical Recipes Chapter 2 begins the "true grit" of numerical analysis by considering the solution of linear algebraic equations. This is done first by Gauss-Jordan elimination (gaussj), and then by LU decomposition with forward and back substitution (ludcmp and lubksb). For singular or nearly singular matrices the best choice is singular value decomposition with back substitution (svdcmp and svbksb). Several linear systems of special form, represented by tridiagonal, Vandermonde, and Toeplitz matrices, may be treated with procedures tridag, vander, and toeplz respectively. Linear systems with relatively few non-zero coefficients, so-called "sparse" matrices, are handled by routine sparse.

$$\star \quad \star \quad \star \quad \star$$

gaussj performs Gauss-Jordan elimination with full pivoting to find the solution of a set of linear equations for a collection of right-hand side vectors. The demonstration routine d2r1 checks its operation with reference to a group of test input matrices printed at the end of this chapter as file matrx1.dat. Each matrix is subjected to inversion by gaussj, and then multiplication by its own inverse to see that a unit matrix is produced. Then the solution vectors are each checked through multiplication by the original matrix and comparison with the right-hand side vectors that produced them.

```
PROGRAM d2r1 (input,output,dfile);
(* driver program for subroutine GAUSSJ *)
(*$I MODFILE.PAS *)
LABEL 99;
CONST
   np = 20;
   mp = 20;
TYPE
   RealArrayNPbyNP = ARRAY [1..np,1..np] OF real;
   RealArrayNPbyMP = ARRAY [1..np,1..mp] OF real;
   IntegerArrayNP = ARRAY [1..np] of integer;
VAR
   j,k,l,m,n: integer;
   a,ai,u: RealArrayNPbyNP;
   b,x,t: RealArrayNPbyMP;
   dfile: text;
(*$I GAUSSJ.PAS *)
BEGIN
   NROpen(dfile,'matrx1.dat');
   WHILE true DO BEGIN
      readln(dfile);
```

```
      readln(dfile);
      readln(dfile,n,m);
      readln(dfile);
      FOR k := 1 TO n DO BEGIN
         FOR l := 1 TO n-1 DO read(dfile,a[k,l]);
         readln(dfile,a[k,n])
      END;
      readln(dfile);
      FOR l := 1 TO m DO BEGIN
         FOR k := 1 TO n-1 DO read(dfile,b[k,l]);
         readln(dfile,b[n,l])
      END;
(* save matrices for later testing of results *)
      FOR l := 1 TO n DO BEGIN
         FOR k := 1 TO n DO ai[k,l] := a[k,l];
         FOR k := 1 TO m DO x[l,k] := b[l,k]
      END;
(* invert matrix *)
      gaussj(ai,n,x,m);
      writeln;
      writeln('Inverse of matrix a : ');
      FOR k := 1 TO n DO BEGIN
         FOR l := 1 TO n-1 DO write(ai[k,l]:12:6);
         writeln(ai[k,n]:12:6)
      END;
(* test results -- check inverse *)
      writeln('a times a-inverse (compare with unit matrix)');
      FOR k := 1 TO n DO BEGIN
         FOR l := 1 TO n DO BEGIN
            u[k,l] := 0.0;
            FOR j := 1 TO n DO
               u[k,l] := u[k,l]+a[k,j]*ai[j,l]
         END;
         FOR l := 1 TO n-1 DO write(u[k,l]:12:6);
         writeln(u[k,n]:12:6)
      END;
(* check vector solutions *)
      writeln;
      writeln('Check the following vectors for equality:');
      writeln('original':20,'matrix*sol''n':15);
      FOR l := 1 TO m DO BEGIN
      writeln('vector ',l:2,':');
         FOR k := 1 TO n DO BEGIN
            t[k,l] := 0.0;
            FOR j := 1 TO n DO
               t[k,l] := t[k,l]+a[k,j]*x[j,l];
            writeln(' ':8,b[k,l]:12:6,t[k,l]:12:6);
         END
      END;
      writeln('********************************');
      IF eof(dfile) THEN GOTO 99;
      writeln('press RETURN for next problem:');
      readln
   END;
99:
   close(dfile)
END.
```

The demonstration program for routine ludcmp relies on the same package of test matrices, but just performs an LU decomposition of each. The performance is checked by multiplying the lower and upper matrices of the decomposition and comparing with the original matrix. The array indx keeps track of the scrambling done by ludcmp to effect partial pivoting. We had to do the unscrambling here, but you will normally not be called upon to do so, since ludcmp is used with the descrambler-containing routine lubksb.

```pascal
PROGRAM d2r2 (input,output,dfile);
(* driver for routine LUDCMP *)
(*$I MODFILE.PAS *)
LABEL 99;
CONST
   np = 20;
TYPE
   RealArrayNPbyNP = ARRAY [1..np,1..np] OF real;
   RealArrayNP = ARRAY [1..np] OF real;
   IntegerArrayNP = ARRAY [1..np] OF integer;
VAR
   j,k,l,m,n,dum: integer;
   d: real;
   a,xl,xu,x: RealArrayNPbyNP;
   indx,jndx: IntegerArrayNP;
   dfile: text;
(*$I LUDCMP.PAS *)
BEGIN
   NROpen(dfile,'matrx1.dat');
   WHILE true DO BEGIN
      readln(dfile);
      readln(dfile);
      readln(dfile,n,m);
      readln(dfile);
      FOR k := 1 TO n DO BEGIN
         FOR l := 1 TO n-1 DO read(dfile,a[k,l]);
         readln(dfile,a[k,n])
      END;
      readln(dfile);
      FOR l := 1 TO m DO BEGIN
         FOR k := 1 TO n-1 DO read(dfile,x[k,l]);
         readln(dfile,x[n,l])
      END;
(* print out a-matrix for comparison with product of lower *)
(* and upper decomposition matrices *)
      writeln('original matrix:');
      FOR k := 1 TO n DO BEGIN
         FOR l := 1 TO n-1 DO write(a[k,l]:12:6);
         writeln(a[k,n]:12:6)
      END;
(* perform the decomposition *)
      ludcmp(a,n,indx,d);
(* compose separately the lower and upper matrices *)
      FOR k := 1 TO n DO BEGIN
         FOR l := 1 TO n DO BEGIN
            IF l > k ThEN BEGIN
               xu[k,l] := a[k,l];
               xl[k,l] := 0.0
```

```
              END ELSE IF 1 < k THEN BEGIN
                 xu[k,1] := 0.0;
                 xl[k,1] := a[k,1]
              END ELSE BEGIN
                 xu[k,1] := a[k,1];
                 xl[k,1] := 1.0
              END
           END
        END;
(* compute product of lower and upper matrices for *)
(* comparison with original matrix *)
     FOR k := 1 TO n DO BEGIN
        jndx[k] := k;
        FOR 1 := 1 TO n DO BEGIN
           x[k,1] := 0.0;
           FOR j := 1 TO n DO
              x[k,1] := x[k,1]+xl[k,j]*xu[j,1]
        END
     END;
     writeln('product of lower and upper matrices (rows unscrambled):');
     FOR k := 1 TO n DO BEGIN
        dum := jndx[indx[k]];
        jndx[indx[k]] := jndx[k];
        jndx[k] := dum
     END;
     FOR k := 1 TO n DO
        FOR j := 1 TO n DO
           IF jndx[j] = k THEN BEGIN
              FOR 1 := 1 TO n-1 DO write(x[j,1]:12:6);
              writeln(x[j,n]:12:6)
           END;
     writeln('lower matrix of the decomposition:');
     FOR k := 1 TO n DO BEGIN
        FOR 1 := 1 TO n-1 DO write(xl[k,1]:12:6);
        writeln(xl[k,n]:12:6)
     END;
     writeln('upper matrix of the decomposition:');
     FOR k := 1 TO n DO BEGIN
        FOR 1 := 1 TO n-1 DO write(xu[k,1]:12:6);
        writeln(xu[k,n]:12:6)
     END;
     writeln('*********************************');
     IF eof(dfile) THEN GOTO 99;
     writeln('press RETURN for next problem:');
     readln
  END;
99:
   close(dfile)
END.
```

Our example driver for lubksb makes calls to both ludcmp and lubksb in order to solve the linear equation problems posed in file matrx1.dat (see discussion of gaussj). The original matrix of coefficients is applied to the solution vectors to check that the result matches the right-hand side vectors posed for each problem. We apologize for using routine ludcmp in a test of lubksb, but ludcmp has been tested independently, and anyway, lubksb is nothing without this partner program, so a test of the combination is more to the point.

```pascal
PROGRAM d2r3(input,output,dfile);
(* driver for routine LUBKSB *)
(*$I MODFILE.PAS *)
LABEL 99;
CONST
   np = 20;
TYPE
   RealArrayNPbyNP = ARRAY [1..np,1..np] OF real;
   RealArrayNP = ARRAY [1..np] OF real;
   IntegerArrayNP = ARRAY [1..np] OF integer;
VAR
   j,k,l,m,n: integer;
   p: real;
   a,b,c: RealArrayNPbyNP;
   indx: IntegerArrayNP;
   x: RealArrayNP;
   dfile: text;
(*$I LUDCMP.PAS *)
(*$I LUBKSB.PAS *)
BEGIN
   NROpen(dfile,'matrx1.dat');
   WHILE true DO BEGIN
      readln(dfile);
      readln(dfile);
      readln(dfile,n,m);
      readln(dfile);
      FOR k := 1 TO n DO BEGIN
         FOR l := 1 TO n-1 DO read(dfile,a[k,l]);
         readln(dfile,a[k,n])
      END;
      readln(dfile);
      FOR l := 1 TO m DO BEGIN
         FOR k := 1 TO n-1 DO read(dfile,b[k,l]);
         readln(dfile,b[n,l])
      END;
(* save matrix a for later testing *)
      FOR l := 1 TO n DO
         FOR k := 1 TO n DO c[k,l] := a[k,l];
(* do lu decomposition *)
      ludcmp(c,n,indx,p);
(* solve equations for each right-hand vector *)
      FOR k := 1 TO m DO BEGIN
         FOR l := 1 TO n DO x[l] := b[l,k];
         lubksb(c,n,indx,x);
(* test results with original matrix *)
         writeln('right-hand side vector:');
         FOR l := 1 TO n-1 DO write(b[l,k]:12:6);
         writeln(b[n,k]:12:6);
         writeln('result of matrix applied to sol''n vector');
         FOR l := 1 TO n DO BEGIN
            b[l,k] := 0.0;
            FOR j := 1 TO n DO
               b[l,k] := b[l,k]+a[l,j]*x[j]
         END;
         FOR l := 1 TO n-1 DO write(b[l,k]:12:6);
```

```
            writeln(b[n,k]:12:6);
            writeln('**********************************')
      END;
      IF eof(dfile) THEN GOTO 99;
      writeln('press RETURN for next problem:');
      readln
   END;
99:
   close(dfile)
END.
```

Procedure `tridag` solves linear equations with coefficients that form a tridiagonal matrix. We provide at the end of this chapter a second file of matrices `matrix2.dat` for the demonstration driver. In all other respects, the demonstration program d2r4 operates in the same fashion as d2r3.

```
PROGRAM d2r4(input,output,dfile);
(* driver for routine TRIDAG *)
(*$I MODFILE.PAS *)
LABEL 99;
CONST
   np = 20;
TYPE
   RealArrayNP = ARRAY [1..np] OF real;
VAR
   k,n: integer;
   diag,superd,subd,rhs,u: RealArrayNP;
   dfile: text;
(*$I TRIDAG.PAS *)
BEGIN
   NROpen(dfile,'matrx2.dat');
   WHILE true DO BEGIN
      readln(dfile);
      readln(dfile);
      readln(dfile,n);
      readln(dfile);
      FOR k := 1 TO n-1 DO read(dfile,diag[k]);
      readln(dfile,diag[n]);
      readln(dfile);
      FOR k := 1 TO n-2 DO read(dfile,superd[k]);
      readln(dfile,superd[n-1]);
      readln(dfile);
      FOR k := 2 TO n-1 DO read(dfile,subd[k]);
      readln(dfile,subd[n]);
      readln(dfile);
      FOR k := 1 TO n-1 DO read(dfile,rhs[k]);
      readln(dfile,rhs[n]);
(* carry out solution *)
      tridag(subd,diag,superd,rhs,u,n);
      writeln('the solution vector is:');
      FOR k := 1 TO n-1 DO write(u[k]:12:6);
      writeln(u[n]:12:6);
(* test solution *)
      writeln('(matrix)*(sol''n vector) should be:');
      FOR k := 1 TO n-1 DO write(rhs[k]:12:6);
      writeln(rhs[n]:12:6);
      writeln('actual result is:');
```

```
        FOR k := 1 TO n DO
          IF k = 1 THEN
              rhs[k] := diag[1]*u[1]+superd[1]*u[2]
          ELSE IF k = n THEN
              rhs[k] := subd[n]*u[n-1]+diag[n]*u[n]
          ELSE
              rhs[k] := subd[k]*u[k-1]+diag[k]*u[k]+superd[k]*u[k+1];
        FOR k := 1 TO n-1 DO write(rhs[k]:12:6);
        writeln(rhs[n]:12:6);
        writeln('********************************');
        IF eof(dfile) THEN GOTO 99;
        writeln('press RETURN for next problem:');
        readln
    END;
99:
    close(dfile)
END.
```

mprove is a short routine for improving the solution vector for a set of linear equations, providing that an LU decomposition has been performed on the matrix of coefficients. Our test of this function is to use ludcmp and lubksb to solve a set of equations specified by the assignment statements at the beginning of the program. The solution vector is then corrupted by the addition of random values to each component. mprove works on the corrupted vector to recover the original.

```
PROGRAM d2r5(input,output);
(* driver for routine MPROVE *)
(*$I MODFILE.PAS *)
CONST
    n = 5;
    np = n;
TYPE
    RealArray55 = ARRAY [1..55] OF real;
    RealArrayNP = ARRAY [1..n] OF real;
    IntegerArrayNP = ARRAY [1..n] OF integer;
    RealArrayNPbyNP = ARRAY [1..np,1..np] OF real;
VAR
    Ran3Inext,Ran3Inextp: integer;
    Ran3Ma: RealArray55;
    d: real;
    i,idum,j: integer;
    a,aa: RealArrayNPbyNP;
    b,x: RealArrayNP;
    indx: IntegerArrayNP;
(*$I RAN3.PAS *)
(*$I LUDCMP.PAS *)
(*$I LUBKSB.PAS *)
(*$I MPROVE.PAS *)
BEGIN
    a[1,1] := 1.0; a[1,2] := 2.0; a[1,3] := 3.0; a[1,4] := 4.0;
    a[1,5] := 5.0; a[2,1] := 2.0; a[2,2] := 3.0; a[2,3] := 4.0;
    a[2,4] := 5.0; a[2,5] := 1.0; a[3,1] := 1.0; a[3,2] := 1.0;
    a[3,3] := 1.0; a[3,4] := 1.0; a[3,5] := 1.0; a[4,1] := 4.0;
    a[4,2] := 5.0; a[4,3] := 1.0; a[4,4] := 2.0; a[4,5] := 3.0;
    a[5,1] := 5.0; a[5,2] := 1.0; a[5,3] := 2.0; a[5,4] := 3.0;
    a[5,5] := 4.0;
```

```
   b[1] := 1.0; b[2] := 1.0; b[3] := 1.0; b[4] := 1.0; b[5] := 1.0;
   FOR i := 1 TO n DO BEGIN
      x[i] := b[i];
      FOR j := 1 TO n DO aa[i,j] := a[i,j]
   END;
   ludcmp(aa,n,indx,d);
   lubksb(aa,n,indx,x);
   writeln;
   writeln('Solution vector for the equations:');
   FOR i := 1 TO n DO write(x[i]:12:6);
   writeln;
(* now phoney up x and let mprove fit it *)
   idum := -13;
   FOR i := 1 TO n DO x[i] := x[i]*(1.0+0.2*ran3(idum));
   writeln;
   writeln('Solution vector with noise added:');
   FOR i := 1 TO n DO write(x[i]:12:6);
   writeln;
   mprove(a,aa,n,indx,b,x);
   writeln;
   writeln('Solution vector recovered by mprove:');
   FOR i := 1 TO n DO write(x[i]:12:6);
   writeln
END.
```

Vandermonde matrices of dimension $N \times N$ have elements that are entirely integer powers of N arbitrary numbers $x_1 \ldots x_N$. (See *Numerical Recipes* for details). In the demonstration program d2r6 we provide five such numbers to specify a 5×5 matrix, and five elements of a right-hand side vector Q. Routine **vander** is used to find the solution vector W. This vector is tested by applying the matrix to W and comparing the result to Q.

```
PROGRAM d2r6(input,output);
(* driver for routine VANDER *)
(*$I MODFILE.PAS *)
CONST
   n = 5;
TYPE
   DoubleArrayNP = ARRAY [1..n] OF double;
VAR
   i,j: integer;
   sum: double;
   q,term,w,x: DoubleArrayNP;
(*$I VANDER.PAS *)
BEGIN
   x[1] := 1.0; x[2] := 1.5; x[3] := 2.0; x[4] := 2.5; x[5] := 3.0;
   q[1] := 1.0; q[2] := 1.5; q[3] := 2.0; q[4] := 2.5; q[5] := 3.0;
   vander(x,w,q,n);
   writeln('Solution vector:');
   FOR i := 1 TO n DO writeln('w[':7,i:1,'] := ',w[i]:12);
   writeln;
   writeln('Test of solution vector:');
   writeln('mtrx*sol''n':14,'original':11);
   sum := 0.0;
   FOR i := 1 TO n DO BEGIN
      term[i] := w[i];
```

```
      sum := sum+w[i]
   END;
   writeln(sum:12:4,q[1]:12:4);
   FOR i := 2 TO n DO BEGIN
      sum := 0.0;
      FOR j := 1 TO n DO BEGIN
         term[j] := term[j]*x[j];
         sum := sum+term[j]
      END;
      writeln(sum:12:4,q[i]:12:4)
   END
END.
```

A very similar test is applied to **toeplz**, which operates on Toeplitz matrices. The $N \times N$ Toeplitz matrix is specified by $2N - 1$ numbers r_i, in this case taken to be simply a linear progression of values. A right-hand side y_i is chosen likewise. **toeplz** finds the solution vector x_i, and checks it in the usual fashion.

```
PROGRAM d2r7(input,output);
(* driver for routine TOEPLZ *)
(*$I MODFILE.PAS *)
CONST
   n = 5;
   twonm1 = 9;      (* twonm1=2*n-1 *)
TYPE
   RealArrayNP = ARRAY [1..n] OF double;
   RealArray2Nm1 = ARRAY [1..twonm1] OF double;
VAR
   i,j: integer;
   sum: double;
   r: RealArray2Nm1;
   x,y: RealArrayNP;
(*$I TOEPLZ.PAS *)
BEGIN
   FOR i := 1 TO n DO y[i] := 0.1*i*i;
   FOR i := 1 TO twonm1 DO r[i] := 0.1*i*i;
   toeplz(r,x,y,n);
   writeln('Solution vector:');
   FOR i := 1 TO n DO writeln('x[':7,i:1,'] :=',x[i]:13);
   writeln;
   writeln('Test of solution:');
   writeln('mtrx*soln':13,'original':12);
   FOR i := 1 TO n DO BEGIN
      sum := 0.0;
      FOR j := 1 TO n DO
         sum := sum+r[n+i-j]*x[j];
      writeln(sum:12:4,y[i]:12:4)
   END
END.
```

The pair **svdcmp**, **svbksb** are tested in the same manner as **ludcmp**, **lubksb**. That is, **svdcmp** is checked independently to see that it yields proper decomposition of matrices. Then the pair of programs is tested as a unit to see that they provide correct solutions to some linear sets. (Note: Because of the order of programs in *Numerical Recipes*, the test of the pair in this case comes first). The matrices and solution vectors are given in the Appendix as file **matrx1.dat**.

Driver d2r8 brings in matrices a and right-hand side vectors b from `matrx1.dat`. Matrix a, itself, is saved for later use. It is copied into matrix u for processing by svdcmp. The results of the processing are the three arrays u, w, v which form the singular value decomposition of a. The right-hand side vectors are fed one at a time to vector c, and the resulting solution vectors x are checked for accuracy through application of the saved matrix a.

```
PROGRAM d2r8(input,output,dfile);
(* driver for routine SVBKSB *)
(*$I MODFILE.PAS *)
LABEL 99;
CONST
   np = 20;
   mp = 20;
TYPE
   RealArrayNP = ARRAY [1..np] OF real;
   RealArrayMP = ARRAY [1..mp] OF real;
   RealArrayNPbyNP = ARRAY [1..np,1..np] OF real;
   RealArrayMPbyNP = ARRAY [1..mp,1..np] OF real;
VAR
   j,k,l,m,n: integer;
   wmax,wmin: real;
   a,b,u: RealArrayMPbyNP;
   v: RealArrayNPbyNP;
   w,x: RealArrayNP;
   c: RealArrayMP;
   dfile: text;
(*$I SVDCMP.PAS *)
(*$I SVBKSB.PAS *)
BEGIN
   NROpen(dfile,'matrx1.dat');
   WHILE true DO BEGIN
      readln(dfile);
      readln(dfile);
      readln(dfile,n,m);
      readln(dfile);
      FOR k := 1 TO n DO BEGIN
         FOR l := 1 TO n-1 DO read(dfile,a[k,l]);
         readln(dfile,a[k,n])
      END;
      readln(dfile);
      FOR l := 1 TO m DO BEGIN
         FOR k := 1 TO n-1 DO read(dfile,b[k,l]);
         readln(dfile,b[n,l])
      END;
(* copy a into u *)
      FOR k := 1 TO n DO
         FOR l := 1 TO n DO
            u[k,l] := a[k,l];
(* decompose matrix a *)
      svdcmp(u,n,n,w,v);
(* find maximum singular value *)
      wmax := 0.0;
      FOR k := 1 TO n DO
         IF w[k] > wmax THEN  wmax := w[k];
(* define "small" *)
      wmin := wmax*(1.0e-6);
```

```
(* zero the "small" singular values *)
      FOR k := 1 TO n DO
            IF w[k] < wmin THEN  w[k] := 0.0;
(* backsubstitute for each right-hand side vector *)
      FOR l := 1 TO m DO BEGIN
            writeln;
            writeln('Vector number ',l:2);
            FOR k := 1 TO n DO c[k] := b[k,l];
            svbksb(u,w,v,n,n,c,x);
            writeln('    solution vector is:');
            FOR k := 1 TO n-1 DO write(x[k]:12:6);
            writeln(x[n]:12:6);
            writeln('    original right-hand side vector:');
            FOR k := 1 TO n-1 DO write(c[k]:12:6);
            writeln(c[n]:12:6);
            writeln('    result of (matrix)*(sol''n vector):');
            FOR k := 1 TO n DO BEGIN
               c[k] := 0.0;
               FOR j := 1 TO n DO c[k] := c[k]+a[k,j]*x[j]
            END;
            FOR k := 1 TO n-1 DO write(c[k]:12:6);
            writeln(c[n]:12:6)
      END;
      writeln('*********************************');
      IF eof(dfile) THEN GOTO 99;
      writeln('press RETURN for next problem');
      readln
   END;
99:
   close(dfile)
END.
```

Companion driver d2r9 takes matrices from matrx3.dat and passes copies u to svdcmp for singular value decomposition into u, w, and v. Then u, w, and the transpose of v are multiplied together. The result is compared to a saved copy of a.

```
PROGRAM d2r9 (input,output,dfile);
(* driver for routine SVDCMP *)
(*$I MODFILE.PAS *)
LABEL 99;
CONST
   np = 20;
   mp = 20;
TYPE
   RealArrayNP = ARRAY [1..np] OF real;
   RealArrayMPbyNP = ARRAY [1..mp,1..np] OF real;
   RealArrayNPbyNP = ARRAY [1..np,1..np] OF real;
VAR
   j,k,l,m,n: integer;
   a,u: RealArrayMPbyNP;
   v: RealArrayNPbyNP;
   w: RealArrayNP;
   dfile: text;
(*$I SVDCMP.PAS *)
BEGIN
(* read input matrices *)
   NROpen(dfile,'matrx3.dat');
```

```
    WHILE true DO BEGIN
       readln(dfile);
       readln(dfile);
       readln(dfile,m,n);
       readln(dfile);
(* copy original matrix into u *)
       FOR k := 1 TO m DO BEGIN
          FOR l := 1 TO n DO BEGIN
             read(dfile,a[k,l]);
             u[k,l] := a[k,l]
          END;
          readln(dfile)
       END;
(* perform decomposition *)
       svdcmp(u,m,n,w,v);
(* write results *)
       writeln('Decomposition matrices:');
       writeln('Matrix u');
       FOR k := 1 TO m DO BEGIN
          FOR l := 1 TO n DO write(u[k,l]:12:6);
          writeln
       END;
       writeln('Diagonal of matrix w');
       FOR k := 1 TO n DO write(w[k]:12:6);
       writeln;
       writeln('Matrix v-transpose');
       FOR k := 1 TO n DO BEGIN
          FOR l := 1 TO n DO write(v[l,k]:12:6);
          writeln
       END;
       writeln;
       writeln('Check product against original matrix:');
       writeln('Original matrix:');
       FOR k := 1 TO m DO BEGIN
          FOR l := 1 TO n DO write(a[k,l]:12:6);
          writeln
       END;
       writeln('Product u*w*(v-transpose):');
       FOR k := 1 TO m DO BEGIN
          FOR l := 1 TO n DO BEGIN
             a[k,l] := 0.0;
             FOR j := 1 TO n DO
                a[k,l] := a[k,l]+u[k,j]*w[j]*v[l,j]
          END;
          FOR l := 1 TO n-1 DO write(a[k,l]:12:6);
          writeln(a[k,n]:12:6);
       END;
       writeln('*********************************');
       IF eof(dfile) THEN GOTO 99;
       writeln('press RETURN for next problem');
       readln
    END;
99:
    close(dfile)
END.
```

Routine **sparse** solves linear systems $\mathbf{A} \cdot \mathbf{x} = \mathbf{b}$ with a sparse matrix $\mathbf{A}$. Rather than specifying the entire matrix **A** (most elements of which are zero), the program calls two procedures `asub` and `atsub` which are, for any input vector `xin`, supposed to return the result `xout` of applying **A** and its transpose to `xin`, respectively. In our sample program we define these two procedures to implement the 20×20 matrix

$$\begin{pmatrix} 1.0 & 2.0 & 0.0 & 0.0 & \cdots \\ -2.0 & 1.0 & 2.0 & 0.0 & \cdots \\ 0.0 & -2.0 & 1.0 & 2.0 & \cdots \\ 0.0 & 0.0 & -2.0 & 1.0 & \cdots \\ \vdots & \vdots & \vdots & \vdots & \ddots \end{pmatrix}$$

As a right-hand side vector **b** we have taken $(3.0, 1.0, 1.0, \ldots, -1.0)$, and the solution is given as **x**. Notice that the components of **x** are all initialized to zero. You will set them to some initial guess of the solution to your own problem, but this guess will usually suffice. The solution in **d2r10** is given the usual checks.

```
PROGRAM d2r10(input,output);
(* driver for routine SPARSE *)
(*$I MODFILE.PAS *)
CONST
    n = 20;
TYPE
    RealArrayNP = ARRAY [1..n] OF real;
VAR
    i,ii: integer;
    rsq: real;
    b,bcmp,x: RealArrayNP;
PROCEDURE asub(VAR xin,xout: RealArrayNP; n: integer);
VAR
    i: integer;
BEGIN
    xout[1] := xin[1]+2.0*xin[2];
    xout[n] := -2.0*xin[n-1]+xin[n];
    FOR i := 2 TO n-1 DO
        xout[i] := -2.0*xin[i-1]+xin[i]+2.0*xin[i+1]
END;
PROCEDURE atsub(VAR xin,xout: RealArrayNP; n: integer);
VAR
    i: integer;
BEGIN
    xout[1] := xin[1]-2.0*xin[2];
    xout[n] := 2.0*xin[n-1]+xin[n];
    FOR i := 2 TO n-1 DO
        xout[i] := 2.0*xin[i-1]+xin[i]-2.0*xin[i+1]
END;
(*$I SPARSE.PAS *)
BEGIN
    FOR i := 1 TO n DO BEGIN
        x[i] := 0.0;
        b[i] := 1.0
    END;
    b[1] := 3.0;
    b[n] := -1.0;
    sparse(b,n,x,rsq);
    writeln('sum-squared residual:',rsq:15);
    writeln;
```

```
     writeln('solution vector:');
     FOR ii := 1 TO n DIV 5 DO BEGIN
        FOR i := 5*(ii-1)+1 TO 5*ii DO write(x[i]:12:6);
        writeln
     END;
     IF n MOD 5 > 0 THEN
        FOR i := 1 TO n MOD 5 DO write(x[5*(n DIV 5)+i]:12:6);
     writeln;
     asub(x,bcmp,n);
     writeln;
     writeln('press RETURN to continue...');
     readln;
     writeln('test of solution vector:');
     writeln('a*x':9,'b':12);
     FOR i := 1 TO n DO writeln(bcmp[i]:12:6,b[i]:12:6)
END.
```

Appendix

File matrx1.dat:

```
MATRICES FOR INPUT TO TEST ROUTINES
Size of matrix (NxN), Number of solutions:
3 2
Matrix A:
1.0 0.0 0.0
0.0 2.0 0.0
0.0 0.0 3.0
Solution vectors:
1.0 0.0 0.0
1.0 1.0 1.0
NEXT PROBLEM
Size of matrix (NxN), Number of solutions:
3 2
Matrix A:
1.0 2.0 3.0
2.0 2.0 3.0
3.0 3.0 3.0
Solution vectors:
1.0 1.0 1.0
1.0 2.0 3.0
NEXT PROBLEM:
Size of matrix (NxN), Number of solutions:
5 2
Matrix A:
1.0 2.0 3.0 4.0 5.0
2.0 3.0 4.0 5.0 1.0
3.0 4.0 5.0 1.0 2.0
4.0 5.0 1.0 2.0 3.0
5.0 1.0 2.0 3.0 4.0
Solution vectors:
1.0 1.0 1.0 1.0 1.0
1.0 2.0 3.0 4.0 5.0
NEXT PROBLEM:
Size of matrix (NxN), Number of solutions:
5 2
Matrix A:
```

```
1.4 2.1 2.1 7.4 9.6
1.6 1.5 1.1 0.7 5.0
3.8 8.0 9.6 5.4 8.8
4.6 8.2 8.4 0.4 8.0
2.6 2.9 0.1 9.6 7.7
Solution vectors:
1.1 1.6 4.7 9.1 0.1
4.0 9.3 8.4 0.4 4.1
```

File matrx2.dat:

```
FILE OF TRIDIAGONAL MATRICES FOR PROGRAM 'TRIDAG'
Dimension of matrix
3
Diagonal elements (N)
1.0 2.0 3.0
Super-diagonal elements (N-1)
2.0 3.0
Sub-diagonal elements (N-1)
2.0 3.0
Right-hand side vector (N)
1.0 2.0 3.0
NEXT PROBLEM:
Dimension of matrix
5
Diagonal elements (N)
1.0 1.0 1.0 1.0 1.0
Super-diagonal elements (N-1)
1.0 2.0 3.0 4.0
Sub-diagonal elements (N-1)
2.0 3.0 4.0 5.0
Right-hand side vector (N)
1.0 2.0 3.0 4.0 5.0
NEXT PROBLEM:
Dimension of matrix
5
Diagonal elements (N)
1.0 2.0 3.0 4.0 5.0
Super-diagonal elements (N-1)
2.0 3.0 4.0 5.0
Sub-diagonal elements (N-1)
2.0 3.0 4.0 5.0
Right-hand side vector (N)
1.0 1.0 1.0 1.0 1.0
NEXT PROBLEM:
Dimension of matrix
6
Diagonal elements (N)
9.7 9.5 5.2 3.5 5.1 6.0
Super-diagonal elements (N-1)
6.0 1.2 0.7 3.0 1.5
Sub-diagonal elements (N-1)
2.1 9.4 3.3 7.5 8.8
Right-hand side vector (N)
2.0 7.5 0.6 7.4 9.8 8.8
```

File `matrx3.dat`:

```
FILE OF MATRICES FOR SVDCMP:
Number of Rows, Columns
5 3
Matrix
1.0 2.0 3.0
2.0 3.0 4.0
3.0 4.0 5.0
4.0 5.0 6.0
5.0 6.0 7.0
NEXT PROBLEM:
Number of Rows, Columns
5 5
Matrix
1.0 2.0 3.0 4.0 5.0
2.0 2.0 3.0 4.0 5.0
3.0 3.0 3.0 4.0 5.0
4.0 4.0 4.0 4.0 5.0
5.0 5.0 5.0 5.0 5.0
NEXT PROBLEM:
Number of Rows, Columns
6 6
Matrix
3.0 5.3 5.6 3.5 6.8 5.7
0.4 8.2 6.7 1.9 2.2 5.3
7.8 8.3 7.7 3.3 1.9 4.8
5.5 8.8 3.0 1.0 5.1 6.4
5.1 5.1 3.6 5.8 5.7 4.9
3.5 2.7 5.7 8.2 9.6 2.9
```

Chapter 3: Interpolation and Extrapolation

Chapter 3 of Numerical Recipes deals with interpolation and extrapolation (the same routines are usable for both). Three fundamental interpolation methods are first discussed,

1. *Polynomial interpolation (`polint`),*

2. *Rational function interpolation (`ratint`), and*

3. *Cubic spline interpolation (`spline`, `splint`).*

To find the place in an ordered table at which to perform an interpolation, two routines are given, `locate` and `hunt`. Also, for cases in which the actual coefficients of a polynomial interpolation are desired, the routines `polcoe` and `polcof` are provided (along with important warnings circumscribing their usefulness).

For higher-dimensional interpolations, Numerical Recipes treats only problems on a regularly spaced grid. Routine `polin2` does a two-dimensional polynomial interpolation that aims at accuracy rather than smoothness. When smooth interpolation is desired, the methods shown in `bcucof` and `bcuint` for bicubic interpolation are recommended. In the case of two-dimensional spline interpolations, the routines `splie2` and `splin2` are offered.

⋆   ⋆   ⋆   ⋆

Program `polint` takes two arrays `xa` and `ya` of length N that express the known values of a function, and calculates the value, at a point x, of the unique polynomial of degree $N - 1$ passing through all the given values. For the purpose of illustration, in d3r1 we have taken evenly spaced `xa[i]` and set `ya[i]` equal to simple functions (sines and exponentials) of these `xa[i]`. For the sine we use an interval of length π, and for the exponential an interval of length 1.0. You may choose the number N of reference points and observe the improvement of the results as N increases. The test points x are slightly shifted from the reference points so that you can compare the estimated error `dy` with the actual error. By removing the shift, you may check that the polynomial actually hits all reference points.

```
PROGRAM d3r1 (input,output);
(* driver for routine POLINT *)
(*$I MODFILE.PAS *)
CONST
   np = 10;   (* maximum value for n *)
   pi = 3.1415926;
TYPE
   RealArrayNP = ARRAY [1..np] OF real;
```

22

```
VAR
    i,n,nfunc: integer;
    dy,f,x,y: real;
    xa,ya: RealArrayNP;
(*$I POLINT.PAS *)
BEGIN
    writeln('generation of interpolation tables');
    writeln(' ... sin(x)   0<x<pi');
    writeln(' ... exp(x)   0<x<1 ');
    writeln('how many entries go in these tables? (note: n<=10)');
    readln(n);
    FOR nfunc := 1 TO 2 DO BEGIN
        writeln;
        IF nfunc = 1 THEN BEGIN
            writeln('sine function from 0 to pi');
            FOR i := 1 TO n DO BEGIN
                xa[i] := i*pi/n;
                ya[i] := sin(xa[i])
            END;
        END ELSE IF nfunc = 2 THEN BEGIN
            writeln('exponential function from 0 to 1');
            FOR i := 1 TO n DO BEGIN
                xa[i] := i*1.0/n;
                ya[i] := exp(xa[i])
            END;
        END;
        writeln;
        writeln('x':9,'f(x)':13,'interpolated':16,'error':11);
        FOR i := 1 TO 10 DO BEGIN
            IF nfunc = 1 THEN BEGIN
                x := (-0.05+i/10.0)*pi;
                f := sin(x)
            END ELSE IF nfunc = 2 THEN BEGIN
                x := (-0.05+i/10.0);
                f := exp(x)
            END;
            polint(xa,ya,n,x,y,dy);
            writeln(x:12:6,f:12:6,y:12:6,' ':4,dy:11)
        END;
        writeln;
        writeln('*********************************');
        writeln('press RETURN');
        readln
    END
END.
```

ratint is functionally similar to polint in that it also returns a value y for the function at point x, and an error estimate dy as well. In this case the values are determined from the unique diagonal rational function that passes through all the reference points. If you inspect the driver closely, you will find that two of the test points fall directly on top of reference points and should give exact results. The remainder do not. You can compare the estimated error dyy to the actual error $|yy - yexp|$ for these cases.

```
PROGRAM d3r2(input,output);
(* driver for routine RATINT *)
(*$I MODFILE.PAS *)
CONST
   npt = 6;
   eps = 1.0;
TYPE
   RealArrayNP = ARRAY [1..npt] OF real;
VAR
   dyy,xx,yexp,yy: real;
   i: integer;
   x,y: RealArrayNP;
FUNCTION f(x,eps: real): real;
BEGIN
   f := x*exp(-x)/(sqr(x-1.0)+eps*eps)
END;
(*$I RATINT.PAS *)
BEGIN
   FOR i := 1 TO npt DO BEGIN
      x[i] := i*2.0/npt;
      y[i] := f(x[i],eps)
   END;
   writeln('Diagonal rational function interpolation');
   writeln;
   writeln('x':5,'interp.':13,'accuracy':14,'actual':12);
   FOR i := 1 TO 10 DO BEGIN
      xx := 0.2*i;
      ratint(x,y,npt,xx,yy,dyy);
      yexp := f(xx,eps);
      writeln(xx:6:2,yy:12:6,' ':4,dyy:11,yexp:12:6)
   END
END.
```

Procedure spline generates a cubic spline. Given an array of x_i and $f(x_i)$, and given values of the first derivative of function f at the two endpoints of the tabulated region, it returns the second derivative of f at each of the tabulation points. As an example we chose the function $\sin x$ and evaluated it at evenly spaced points x[i]. In this case the first derivatives at the end-points are yp1 $= \cos x_1$ and ypn $= \cos x_N$. The output array of spline is y2[i] and this is listed along with $-\sin x_i$, the second derivative of $\sin x_i$, for comparison.

```
PROGRAM d3r3 (input,output);
(* driver for routine SPLINE *)
(*$I MODFILE.PAS *)
CONST
   n = 20;
   pi = 3.1415926;
TYPE
   RealArrayNP = ARRAY [1..n] OF real;
VAR
   i: integer;
   yp1,ypn: real;
   x,y,y2: RealArrayNP;
(*$I SPLINE.PAS *)
BEGIN
   writeln('second-derivatives for sin(x) from 0 to pi');
```

```
(* generate array for interpolation *)
  FOR i := 1 TO 20 DO BEGIN
    x[i] := i*pi/n;
    y[i] := sin(x[i])
  END;
(* calculate 2nd derivative with spline *)
  yp1 := cos(x[1]);
  ypn := cos(x[n]);
  spline(x,y,n,yp1,ypn,y2);
(* test result *)
  writeln('spline':23,'actual':16);
  writeln('number':11,'2nd deriv':14,'2nd deriv':16);
  FOR i := 1 TO n DO writeln(i:8,y2[i]:16:6,-sin(x[i]):16:6)
END.
```

Actual cubic-spline interpolations, however, are carried out by **splint**. This routine uses the output array from one call to **spline** to service any subsequent number of spline interpolations with different x's. The demonstration program d3r4 tests this capability on both $\sin x$ and $\exp x$. The two are treated in succession according to whether nfunc is one or two. In each case the function is tabulated at equally spaced points, and the derivatives are found at the first and last point. A call to **spline** then produces an array of second derivatives y2 which is fed to **splint**. The interpolated values y are compared with actual function values f at a different set of equally spaced points.

```
PROGRAM d3r4 (input,output);
(* driver for routine SPLINT *)
(*$I MODFILE.PAS *)
CONST
   np = 10;
   pi = 3.1415926;
TYPE
   RealArrayNP = ARRAY [1..np] OF real;
VAR
   i,nfunc: integer;
   f,x,y,yp1,ypn: real;
   xa,ya,y2: RealArrayNP;
(*$I SPLINE.PAS *)
(*$I SPLINT.PAS *)
BEGIN
   FOR nfunc := 1 TO 2 DO BEGIN
      writeln;
      IF nfunc = 1 THEN BEGIN
         writeln('sine function from 0 to pi');
         FOR i := 1 TO np DO BEGIN
            xa[i] := i*pi/np;
            ya[i] := sin(xa[i])
         END;
         yp1 := cos(xa[1]);
         ypn := cos(xa[np])
      END ELSE IF nfunc = 2 THEN BEGIN
         writeln('exponential function from 0 to 1');
         FOR i := 1 TO np DO BEGIN
            xa[i] := 1.0*i/np;
            ya[i] := exp(xa[i])
         END;
```

```
        yp1 := exp(xa[1]);
        ypn := exp(xa[np])
    END;
(* call spline to get second derivatives *)
    spline(xa,ya,np,yp1,ypn,y2);
(* call splint for interpolations *)
    writeln;
    writeln('x':9,'f(x)':13,'interpolation':17);
    FOR i := 1 TO 10 DO BEGIN
      IF nfunc = 1 THEN BEGIN
        x := (-0.05+i/10.0)*pi;
        f := sin(x)
      END ELSE IF nfunc = 2 THEN BEGIN
        x := -0.05+i/10.0;
        f := exp(x)
      END;
      splint(xa,ya,y2,np,x,y);
      writeln(x:12:6,f:12:6,y:12:6)
    END;
    writeln;
    writeln('*********************************');
    writeln('press RETURN');
    readln
  END
END.
```

The next program, `locate`, may be used in conjunction with any interpolation method to bracket the x-position for which $f(x)$ is sought by two adjacent tabulated positions. That is, given a monotonic array of x_i, and given a value of x, it finds the two values x_i, x_{i+1} that surround x. In d3r5 we chose the array x_i to be non-uniform, varying exponentially with i. Then we took a uniform series of x-values and sought their position in the array using `locate`. For each x, `locate` finds the value j for which x[j] is nearest below x. Then the driver shows j, and the two bracketing values xx[j] and xx[j+1]. If j is 0 or n, then x is not within the tabulated range. The program thereby flags 'lower lim' if x is below x[1] or 'upper lim' if x is above x[n].

```
PROGRAM d3r5 (input,output);
(* driver for routine LOCATE *)
(*$I MODFILE.PAS *)
CONST
  n = 100;
TYPE
  RealArrayNP = ARRAY [1..n] OF real;
VAR
  i,j: integer;
  x: real;
  xx: RealArrayNP;
(*$I LOCATE.PAS *)
BEGIN
(* create array to be searched *)
  FOR i := 1 TO n DO xx[i] := exp(i/20.0)-74.0;
  writeln('result of:  j := 0 indicates x too small');
  writeln(' ':12,'j := 100 indicates x too large');
  writeln;
  writeln('locate ':10,'j':6,'xx(j)':11,'xx(j+1)':12);
```

```
(* do test *)
   FOR i := 1 TO 19 DO BEGIN
      x := -100.0+200.0*i/20.0;
      locate(xx,n,x,j);
      IF (j < n) AND (j > 0) THEN
         writeln(x:10:4,j:6,xx[j]:12:6,xx[j+1]:12:6)
      ELSE IF j = n THEN
         writeln(x:10:4,j:6,xx[j]:12:6,'   upper lim')
      ELSE
         writeln(x:10:4,j:6,'   lower lim',xx[j+1]:12:6)
   END
END.
```

Routine hunt serves the same function as locate, but is used when the table is to be searched many times and the abscissa each time is close to its value on the previous search. d3r6 sets up the array xx[i] and then a series x of points to locate. The hunt begins with a trial value ji (which is fed to hunt through variable j) and hunt returns solution j such that x lies between xx[j] and xx[j+1]. The two cases j=0 and j=n have the same meaning as in d3r5 and are treated in the same way.

```
PROGRAM d3r6 (input,output);
(* driver for routine HUNT *)
(*$I MODFILE.PAS *)
CONST
   n = 100;
TYPE
   RealArrayNP = ARRAY [1..n] OF real;
VAR
   i,j,ji: integer;
   x: real;
   xx: RealArrayNP;
(*$I HUNT.PAS *)
BEGIN
(* create array to be searched *)
   FOR i := 1 TO n DO xx[i] := exp(i/20.0)-74.0;
   writeln('  result of:',' ':3,'j := 0 indicates x too small');
   writeln(' ':15,'j := 100 indicates x too large');
   writeln('locate:':12,'guess':8,'j':4,'xx(j)':11,'xx(j+1)':13);
(* do test *)
   FOR i := 1 TO 19 DO BEGIN
      x := -100.0+200.0*i/20.0;
(* trial parameter *)
      ji := 5*i;
      j := ji;
(* begin search *)
      hunt(xx,n,x,j);
      IF (j < n) AND (j > 0) THEN
         writeln(x:12:5,ji:6,j:6,xx[j]:12:6,xx[j+1]:12:6)
      ELSE IF j = n THEN
         writeln(x:12:5,ji:6,j:6,xx[j]:12:6,'   upper lim')
      ELSE
         writeln(x:12:5,ji:6,j:6,'   lower lim',xx[j+1]:12:6)
   END
END.
```

The next two demonstration programs, d3r7 and d3r8, are so nearly identical

that they may be discussed together. polcoe and polcof themselves both find coefficients of interpolating polynomials. In the present instance we have tried both a sine
function and an exponential function for ya[i], each tabulated at uniformly spaced
points xa[i]. The validity of the array of polynomial coefficients coeff is tested by
calculating the value sum of the polynomials at a series of test points and listing these
alongside the functions f that they represent.

```pascal
PROGRAM d3r7 (input,output);
(* driver for routine POLCOE *)
(*$I MODFILE.PAS *)
CONST
   np = 5;
   pi = 3.1415926;
TYPE
   RealArrayNP = ARRAY [1..np] OF real;
VAR
   i,j,nfunc: integer;
   f,sum,x: real;
   coeff,xa,ya: RealArrayNP;
(*$I POLCOE.PAS *)
BEGIN
   FOR nfunc := 1 TO 2 DO BEGIN
      IF nfunc = 1 THEN BEGIN
         writeln('sine function from 0 to pi');
         writeln;
         FOR i := 1 TO np DO BEGIN
            xa[i] := i*pi/np;
            ya[i] := sin(xa[i])
         END
      END ELSE IF nfunc = 2 THEN BEGIN
         writeln('exponential function from 0 to 1');
         writeln;
         FOR i := 1 TO np DO BEGIN
            xa[i] := 1.0*i/np;
            ya[i] := exp(xa[i])
         END
      END;
      polcoe(xa,ya,np,coeff);
      writeln(' ':2,'coefficients');
      FOR i := 1 TO np DO writeln(coeff[i]:12:6);
      writeln;
      writeln('x':9,'f(x)':13,'polynomial':15);
      FOR i := 1 TO 10 DO BEGIN
         IF nfunc = 1 THEN BEGIN
            x := (-0.05+i/10.0)*pi;
            f := sin(x)
         END ELSE IF nfunc = 2 THEN BEGIN
            x := -0.05+i/10.0;
            f := exp(x)
         END;
         sum := coeff[np];
         FOR j := np-1 DOWNTO 1 DO sum := coeff[j]+sum*x;
         writeln(x:12:6,f:12:6,sum:12:6)
      END;
      writeln;
      writeln('**********************************');
      writeln('press RETURN');
```

```
          readln
     END
END.


PROGRAM d3r8 (input,output);
(* driver for routine POLCOF *)
(*$I MODFILE.PAS *)
CONST
     np = 5;
     pi = 3.1415926;
TYPE
     RealArrayNP = ARRAY [1..np] OF real;
VAR
     i,j,nfunc: integer;
     f,sum,x: real;
     coeff,xa,ya: RealArrayNP;
(*$I POLINT.PAS *)
(*$I POLCOF.PAS *)
BEGIN
     FOR nfunc := 1 TO 2 DO BEGIN
        IF nfunc = 1 THEN BEGIN
           writeln;
           writeln('sine function from 0 to pi');
           FOR i := 1 TO np DO BEGIN
              xa[i] := i*pi/np;
              ya[i] := sin(xa[i])
           END
        END ELSE IF nfunc = 2 THEN BEGIN
           writeln;
           writeln('exponential function from 0 to 1');
           FOR i := 1 TO np DO BEGIN
              xa[i] := 1.0*i/np;
              ya[i] := exp(xa[i])
           END
        END;
        polcof(xa,ya,np,coeff);
        writeln;
        writeln('  coefficients');
        FOR i := 1 TO np DO writeln(coeff[i]:12:6);
        writeln;
        writeln('x':9,'f(x)':13,'polynomial':15);
        FOR i := 1 TO 10 DO BEGIN
           IF nfunc = 1 THEN BEGIN
              x := (-0.05+i/10.0)*pi;
              f := sin(x)
           END ELSE IF nfunc = 2 THEN BEGIN
              x := -0.05+i/10.0;
              f := exp(x)
           END;
           sum := coeff[np];
           FOR j := np-1 DOWNTO 1 DO sum := coeff[j]+sum*x;
           writeln(x:12:6,f:12:6,sum:12:6)
        END;
        writeln('*********************************');
        writeln('press RETURN');
        readln
     END
```

```
END.
```

For two-dimensional interpolation, polin2 implements a bilinear interpolation. We feed it coordinates x1a,x2a for an $M \times N$ array of gridpoints as well as the function value at each gridpoint. In return it gives the value y of the interpolated function at a given point x1,x2, and the estimated accuracy dy of the interpolation. d3r9 runs the test on a uniform grid for the function $f(x,y) = \sin x \exp y$. Then, for an offset grid of test points, the interpolated value y is compared to the actual function value f, and the actual error is compared to the estimated error dy.

```
PROGRAM d3r9 (input,output);
(* driver for routine POLIN2 *)
(*$I MODFILE.PAS *)
CONST
   n = 5;
   pi = 3.1415926;
TYPE
   RealArrayMP = ARRAY [1..n] OF real;
   RealArrayNP = RealArrayMP;
   RealArrayMPbyNP = ARRAY [1..n,1..n] OF real;
VAR
   i,j: integer;
   dy,f,x1,x2,y: real;
   x1a,x2a: RealArrayMP;
   ya: RealArrayMPbyNP;
(*$I POLINT.PAS *)
(*$I POLIN2.PAS *)
BEGIN
   FOR i := 1 TO n DO BEGIN
      x1a[i] := i*pi/n;
      FOR j := 1 TO n DO BEGIN
         x2a[j] := 1.0*j/n;
         ya[i,j] := sin(x1a[i])*exp(x2a[j])
      END
   END;
(* test 2-dimensional interpolation *)
   writeln;
   writeln('Two dimensional interpolation of sin(x1)exp(x2)');
   writeln;
   writeln('x1':9,'x2':12,'f(x)':13,'interpolated':16,'error':11);
   FOR i := 1 TO 4 DO BEGIN
      x1 := (-0.1+i/5.0)*pi;
      FOR j := 1 TO 4 DO BEGIN
         x2 := -0.1+j/5.0;
         f := sin(x1)*exp(x2);
         polin2(x1a,x2a,ya,n,n,x1,x2,y,dy);
         writeln(x1:12:6,x2:12:6,f:12:6,y:12:6,dy:15:6)
      END;
      writeln('*********************************')
   END
END.
```

Bicubic interpolation in two dimensions is carried out with bcucof and bcuint. The first supplies interpolating coefficients within a grid square and the second calculates interpolated values. The calculation provides not only interpolated function values, but also interpolated values of two partial derivatives; all of which are guaranteed to be smooth. To get this, we are required to supply more information than we have needed in previous interpolation routines.

Demonstration program d3r10 works with the function $f(x,y) = xy\exp(-xy)$. You may compare the two first derivatives and the cross derivative of this function with what you find computed in the routine. The function and derivatives are calculated at the four corners of a rectangular grid cell, in this case a 2×2 unit square with one corner at the origin. The points are supplied counterclockwise around the cell. d1 and d2 are the dimensions of the cell. A call to bcucof provides sixteen coefficients which are listed below for your reference.

```
Coefficients for bicubic interpolation:
    0.000000     0.000000     0.000000     0.000000
    0.000000     4.000000     0.000000     0.000000
    0.000000     0.000000   -13.655598     6.095174
    0.000000     0.000000     6.095174    -2.461486
```

```pascal
PROGRAM d3r10(input,output,infile);
(* driver for routine BCUCOF *)
(*$I MODFILE.PAS *)
TYPE
   RealArray4 = ARRAY [1..4] OF real;
   RealArray4by4  = ARRAY [1..4,1..4] OF real;
   RealArray16by16 = ARRAY [1..16,1..16] OF real;
VAR
   infile: text;
   d1,d2,ee,x1x2: real;
   i,j: integer;
   y,y1,y2,y12,x1,x2: RealArray4;
   c: RealArray4by4;
   BcucofWt: RealArray16by16;
   BcucofFlag: boolean;
(*$I BCUCOF.PAS *)
BEGIN
   BcucofFlag := true;
   x1[1] := 0.0; x1[2] := 2.0; x1[3] := 2.0; x1[4] := 0.0;
   x2[1] := 0.0; x2[2] := 0.0; x2[3] := 2.0; x2[4] := 2.0;
   d1 := x1[2]-x1[1];
   d2 := x2[4]-x2[1];
   FOR i := 1 TO 4 DO BEGIN
      x1x2 := x1[i]*x2[i];
      ee := exp(-x1x2);
      y[i] := x1x2*ee;
      y1[i] := x2[i]*(1.0-x1x2)*ee;
      y2[i] := x1[i]*(1.0-x1x2)*ee;
      y12[i] := (1.0-3.0*x1x2+sqr(x1x2))*ee
   END;
   bcucof(y,y1,y2,y12,d1,d2,c);
   writeln;
   writeln('Coefficients for bicubic interpolation:');
   writeln;
   FOR i := 1 TO 4 DO BEGIN
      FOR j := 1 TO 4 DO write(c[i,j]:12:6);
      writeln
   END
END.
```

Program d3r11 works with the function $f(x, y) = (xy)^2$, which has derivatives $\partial f/\partial x = 2xy^2$, $\partial f/\partial y = 2yx^2$, and $\partial^2 f/\partial x \partial y = 4xy$. These are supplied to bcuint along with the locations of the grid points. bcuint calls bcucof internally to determine coefficients, and then calculates ansy, ansy1, ansy2, the interpolated values of f, $\partial f/\partial x$ and $\partial f/\partial y$ at the specified test point (x1,x2). These are compared by the demonstration program to expected values for the three quantities, which are called ey, ey1, and ey2. The test points run along the diagonal of the grid square.

```
PROGRAM d3r11(input,output,infile);
(* driver for routine BCUINT *)
(*$I MODFILE.PAS *)
TYPE
   RealArray4 = ARRAY [1..4] OF real;
   RealArray4by4 = ARRAY [1..4,1..4] OF real;
   RealArray16by16 = ARRAY [1..16,1..16] OF real;
VAR
   infile: text;
   ansy,ansy1,ansy2,ey,ey1,ey2: real;
   x1,x1l,x1u,x1x2,x2,x2l,x2u,xxyy: real;
   i: integer;
   xx,y,y1,y12,y2,yy: RealArray4;
   BcucofWt: RealArray16by16;
   BcucofFlag: boolean;
(*$I BCUCOF.PAS *)
(*$I BCUINT.PAS *)
BEGIN
   BcucofFlag := true;
   xx[1] := 0.0; xx[2] := 2.0; xx[3] := 2.0; xx[4] := 0.0;
   yy[1] := 0.0; yy[2] := 0.0; yy[3] := 2.0; yy[4] := 2.0;
   x1l := xx[1];
   x1u := xx[2];
   x2l := yy[1];
   x2u := yy[4];
   FOR i := 1 TO 4 DO BEGIN
      xxyy := xx[i]*yy[i];
      y[i] := sqr(xxyy);
      y1[i] := 2.0*yy[i]*xxyy;
      y2[i] := 2.0*xx[i]*xxyy;
      y12[i] := 4.0*xxyy
   END;
   writeln;
   writeln('x1':6,'x2':8,'y':7,'expect':11,'y1':6,
      'expect':10,'y2':6,'expect':10);
   writeln;
   FOR i := 1 TO 10 DO BEGIN
      x1 := 0.2*i;
      x2 := x1;
      bcuint(y,y1,y2,y12,x1l,x1u,x2l,x2u,x1,x2,ansy,ansy1,ansy2);
      x1x2 := x1*x2;
      ey := sqr(x1x2);
      ey1 := 2.0*x2*x1x2;
      ey2 := 2.0*x1*x1x2;
      writeln(x1:8:4,x2:8:4,ansy:8:4,ey:8:4,
         ansy1:8:4,ey1:8:4,ansy2:8:4,ey2:8:4)
   END
END.
```

Routines splie2 and splin2 work as a pair to perform bicubic spline interpolations. splie2 takes a function tabulated on an $M \times N$ grid and performs one dimensional natural cubic splines along the rows of the grid to generate an array of second derivatives. These are fodder for splin2 which takes the grid points, function values, and second derivative values and returns the interpolated function value for a desired point in the grid region.

Demonstration program d3r12 exercises splie2 on a regular 10×10 grid of points with coordinates x1 and x2, for the function $y = (x_1 x_2)^2$. The calculated second derivative array is compared with the actual second derivative $2x_1^2$ of the function. Keep in mind that a natural spline is assumed, so that agreement will not be so good near the boundaries of the grid. (This shows that you should *not* assume a natural spline if you have better derivative information at the endpoints.)

```
PROGRAM d3r12(input,output);
(* driver for routine SPLIE2 *)
(*$I MODFILE.PAS *)
CONST
   m = 10;
   n = 10;
TYPE
   RealArrayNP = ARRAY [1..n] OF real;
   RealArrayMPbyNP = ARRAY [1..m,1..n] OF real;
   RealArrayNN = RealArrayNP;
VAR
   i,j: integer;
   x1x2: real;
   x1,x2: RealArrayNP;
   y,y2: RealArrayMPbyNP;
(*$I SPLINE.PAS *)
(*$I SPLIE2.PAS *)
BEGIN
   FOR i := 1 TO m DO x1[i] := 0.2*i;
   FOR i := 1 TO n DO x2[i] := 0.2*i;
   FOR i := 1 TO m DO
      FOR j := 1 TO n DO BEGIN
         x1x2 := x1[i]*x2[j];
         y[i,j] := sqr(x1x2)
      END;
   splie2(x1,x2,y,m,n,y2);
   writeln;
   writeln('second derivatives from SPLIE2');
   writeln('natural spline assumed');
   FOR i := 1 TO 5 DO BEGIN
      FOR j := 1 TO 5 DO write(y2[i,j]:12:6);
      writeln
   END;
   writeln;
   writeln('actual second derivatives');
   FOR i := 1 TO 5 DO BEGIN
      FOR j := 1 TO 5 DO
         y2[i,j] := 2.0*sqr(x1[i]);
      FOR j := 1 TO 5 DO write(y2[i,j]:12:6);
      writeln
   END
END.
```

The demonstration program d3r13 establishes a similar 10×10 grid for the function $y = x_1 x_2 \exp(-x_1 x_2)$. It makes a single call to splie2 to produce second derivatives y2, and then finds function values f through calls to splin2, comparing them to actual function values ff. These values are determined, for no reason better than perversity, along a quadratic path $x_2 = x_1^2$ through the grid region.

```
PROGRAM d3r13(input,output);
(* driver for routine SPLIN2 *)
(*$I MODFILE.PAS *)
CONST
   m = 10;
   n = 10;
TYPE
   RealArrayNP = ARRAY [1..n] OF real;
   RealArrayMPbyNP = ARRAY [1..m,1..n] OF real;
   RealArrayNN = RealArrayNP;
VAR
   f,ff,x1x2,xx1,xx2: real;
   i,j: integer;
   x1,x2: RealArrayNP;
   y,y2: RealArrayMPbyNP;
(*$I SPLINT.PAS *)
(*$I SPLINE.PAS *)
(*$I SPLIE2.PAS *)
(*$I SPLIN2.PAS *)
BEGIN
   FOR i := 1 TO m DO x1[i] := 0.2*i;
   FOR i := 1 TO n DO x2[i] := 0.2*i;
   FOR i := 1 TO m DO
      FOR j := 1 TO n DO BEGIN
         x1x2 := x1[i]*x2[j];
         y[i,j] := x1x2*exp(-x1x2)
      END;
   splie2(x1,x2,y,m,n,y2);
   writeln('x1':9,'x2':12,'splin2':14,'actual':12);
   FOR i := 1 TO 10 DO BEGIN
      xx1 := 0.1*i;
      xx2 := sqr(xx1);
      splin2(x1,x2,y,y2,m,n,xx1,xx2,f);
      x1x2 := xx1*xx2;
      ff := x1x2*exp(-x1x2);
      writeln(xx1:12:6,xx2:12:6,f:12:6,ff:12:6)
   END
END.
```

Chapter 4: Integration of Functions

Numerical integration, or "quadrature", has been treated with some degree of detail in *Numerical Recipes* . *Chapter 4 begins with* trapzd, *a procedure for applying the extended trapezoidal rule. It can be used in successive calls for sequentially improving accuracy, and is used as a foundation for several other programs. For example* qtrap *is an integrating routine that makes repeated calls to* trapzd *until a certain fractional accuracy is achieved.* qsimp *also calls* trapzd, *and in this case performs integration by Simpson's rule. Romberg integration, a generalization of Simpson's rule to successively higher orders, is performed with* qromb—*this one also calls* trapzd. *For improper integrals a different "workhorse" is used, the procedure* midpnt. *This routine applies the extended midpoint rule to avoid function evaluations at an endpoint of the region of integration. It can be used in* qtrap *or* qsimp *in place of* trapzd. *Routine* qromb *can be generalized similarly, and we have implemented this idea in* qromo, *a Romberg integrator for open intervals. The chapter also offers a number of exact replacements for* midpnt, *to be used for various types of singularity in the integrand:*

1. midinf – *if one or the other of the limits of integration is infinite.*

2. midsql - *if there is an inverse square root singularity of the integrand at the lower limit of integration.*

3. midsqu - *if there is an inverse square root singularity of the integrand at the upper limit of integration.*

4. midexp - *when the upper limit of integration is infinite and the integrand decreases exponentially at infinity.*

The somewhat more subtle method of Gaussian quadrature uses unequally spaced abscissas, and weighting coefficients which can be read from tables. Routine qgaus *computes integrals with a ten-point Gauss-Legendre weighting using such coefficients.* gauleg *calculates the tables of abscissas and weights that would apply to an N-point Gauss-Legendre quadrature.*

$$\star \quad \star \quad \star \quad \star$$

trapzd applies the extended trapezoidal rule for integration. It is called sequentially for higher and higher stages of refinement of the integral. The sample program d4r1 uses trapzd to perform a numerical integration of the function

$$\text{func} = x^2(x^2 - 2)\sin x$$

whose definite integral is

$$\mathtt{fint} = 4x(x^2 - 7)\sin x - (x^4 - 14x^2 + 28)\cos x.$$

The integral is performed from $A = 0.0$ to $B = \pi/2$. To demonstrate the increasing accuracy on sequential calls, `trapzd` is called 14 times with the index i increasing by one each time. The improving values of the integral are listed for comparison to the actual value $\mathtt{fint}(B) - \mathtt{fint}(A)$.

```
PROGRAM d4r1(input,output);
(* driver for routine trapzd *)
(*$I MODFILE.PAS *)
CONST
   nmax = 14;
   pio2 = 1.5707963;
VAR
   TrapzdIt: integer;
   i: integer;
   a,b,s: real;
FUNCTION func(x: real): real;
(* Test function *)
BEGIN
   func := sqr(x)*(sqr(x)-2.0)*sin(x)
END;
FUNCTION fint(x: real): real;
(* Integral of test function *)
BEGIN
   fint := 4.0*x*(sqr(x)-7.0)*sin(x)-
      (sqr(sqr(x))-14.0*sqr(x)+28.0)*cos(x);
END;
(*$I TRAPZD.PAS *)
BEGIN
   a := 0.0;
   b := pio2;
   writeln('integral of func with 2^(n-1) points');
   writeln('actual value of integral is',fint(b)-fint(a):12:6);
   writeln('n':6,'approx. integral':24);
   FOR i := 1 TO nmax DO BEGIN
      trapzd(a,b,s,i);
      writeln(i:6,s:20:6)
   END
END.
```

`qtrap` carries out the same integration algorithm but allows us to specify the accuracy with which we wish the integration done. (It is specified within `qtrap` as eps=1.0e-6.) `qtrap` itself makes the sequential calls to `trapzd` until the desired accuracy is reached. Then `qtrap` issues a single result. In sample program `d4r2` we compare this result to the exact value of the integral.

```
PROGRAM d4r2(input,output);
(* driver for routine QTRAP *)
(*$I MODFILE.PAS *)
CONST
   pio2 = 1.5707963;
VAR
   TrapzdIt: integer;
   a,b,s: real;
```

```
FUNCTION func(x: real): real;
(* Test function *)
BEGIN
   func := sqr(x)*(sqr(x)-2.0)*sin(x)
END;
FUNCTION fint(x: real): real;
(* Integral of test function *)
BEGIN
   fint := 4.0*x*(sqr(x)-7.0)*sin(x)-
           (sqr(sqr(x))-14.0*sqr(x)+28.0)*cos(x);
END;
(*$I TRAPZD.PAS *)
(*$I QTRAP.PAS *)
BEGIN
   a := 0.0;
   b := pio2;
   writeln('Integral of func computed with QTRAP');
   writeln;
   writeln('Actual value of integral is',fint(b)-fint(a):12:6);
   qtrap(a,b,s);
   writeln('Result from routine QTRAP is',s:12:6);
END.
```

Alternatively, the integral may be handled by qsimp which applies Simpson's rule. Sample program d4r3 carries out the same integration as the previous program, and reports the result in the same way as well.

```
PROGRAM d4r3(input,output);
(* driver for routine QSIMP *)
(*$I MODFILE.PAS *)
CONST
   pio2 = 1.5707963;
VAR
   TrapzdIt: integer;
   a,b,s: real;
FUNCTION func(x: real): real;
(* Test function *)
BEGIN
   func := sqr(x)*(sqr(x)-2.0)*sin(x)
END;
FUNCTION fint(x: real): real;
(* Integral of test function *)
BEGIN
   fint := 4.0*x*(sqr(x)-7.0)*sin(x)-
           (sqr(sqr(x))-14.0*sqr(x)+28.0)*cos(x);
END;
(*$I TRAPZD.PAS *)
(*$I QSIMP.PAS *)
BEGIN
   a := 0.0;
   b := pio2;
   writeln('Integral of func computed with QSIMP');
   writeln;
   writeln('Actual value of integral is',fint(b)-fint(a):12:6);
   qsimp(a,b,s);
   writeln('Result from routine QSIMP is',s:12:6)
END.
```

qromb generalizes Simpson's rule to higher orders. It makes successive calls to trapzd and stores the results. Then it uses polint, the polynomial interpolater/extrapolator, to project the value of integral which would be obtained were we to continue indefinitely with trapzd. Sample program d4r4 is essentially identical to the sample programs for qtrap and qsimp.

```pascal
PROGRAM d4r4(input,output);
(* driver for routine QROMB *)
(*$I MODFILE.PAS *)
CONST
   pio2 = 1.5707963;
   n = 5;
TYPE
   RealArrayNP = ARRAY [1..n] OF real;
VAR
   TrapzdIt: integer;
   a,b,s: real;
FUNCTION func(x: real): real;
(* Test function *)
BEGIN
   func := sqr(x)*(sqr(x)-2.0)*sin(x)
END;
FUNCTION fint(x: real): real;
(* Integral of test function func *)
BEGIN
   fint := 4.0*x*(sqr(x)-7.0)*sin(x)
      -(sqr(sqr(x))-14.0*sqr(x)+28.0)*cos(x)
END;
(*$I TRAPZD.PAS *)
(*$I POLINT.PAS *)
(*$I QROMB.PAS *)
BEGIN
   a := 0.0;
   b := pio2;
   writeln('Integral of func computed with QROMB');
   writeln;
   writeln('Actual value of integral is',fint(b)-fint(a):12:6);
   qromb(a,b,s);
   writeln('Result from routine QROMB is',s:12:6);
END.
```

Sample program d4r5 uses the function func $= 1/\sqrt{x}$ which is singular at the origin. Limits of integration are set at $A = 0.0$ and $B = 1.0$. midpnt, however, implements an open formula and does not evaluate the function exactly at $x = 0$. In this case the integral is compared to $\text{fint}(B) - \text{fint}(A)$ where fint $= 2\sqrt{x}$, the integral of func.

```pascal
PROGRAM d4r5(input,output);
(* driver for routine MIDPNT *)
(*$I MODFILE.PAS *)
CONST
   nmax = 10;
VAR
   a,b,s: real;
   i,MidpntIt: integer;
FUNCTION func(x: real): real;
```

```
(* Function for testing integration *)
BEGIN
   func := 1.0/sqrt(x)
END;
FUNCTION fint(x: real): real;
(* Integral of 'func' *)
BEGIN
   fint := 2.0*sqrt(x)
END;
(*$I MIDPNT.PAS *)
BEGIN
   a := 0.0;
   b := 1.0;
   writeln;
   writeln('Integral of func computed with MIDPNT');
   writeln('Actual value of integral is',(fint(b)-fint(a)):7:4);
   writeln('n':6,'Approx. integral':29);
   FOR i := 1 TO nmax DO BEGIN
      midpnt(a,b,s,i);
      writeln(i:6,s:24:6)
   END
END.
```

One of the special forms of `midpnt`, namely `midsql`, is demonstrated by sample program d4r6. We evaluate the integral of $\sqrt{x}/\sin x$ from 0.0 to $\pi/2$. This has a $1/\sqrt{x}$ singularity at $x = 0$.

```
PROGRAM d4r6(input,output);
(* Driver for routine QROMO *)
(*$I MODFILE.PAS *)
CONST
   x1 = 0.0;
   x2 = 1.5707963;
   n = 5;
TYPE
   RealArrayNP = ARRAY [1..n] OF real;
VAR
   MidsqlIt: integer;
   result: real;
FUNCTION func(x: real): real;
BEGIN
   func := sqrt(x)/sin(x)
END;
(*$I MIDSQL.PAS *)
(*$I POLINT.PAS *)
(*$I QROMO.PAS *)
BEGIN
   writeln('Improper integral:');
   writeln;
   qromo(x1,x2,result);
   writeln('Function: sqrt(x)/sin(x)      Interval: (0,pi/2)');
   writeln('Using: MIDSQL                 Result:',result:8:4);
END.
```

Procedure `qgaus` performs a Gauss-Legendre integration, using only ten function evaluations. Sample program d4r7 applies it to the function $x \exp(-x)$ whose integral from x_1 to x is $(1 + x_1)\exp(-x_1) - (1 + x)\exp(-x)$. qgaus returns this integral as parameter ss. The method is used for a series of intervals, as short as $(0.0 - 0.5)$ and as long as $(0.0 - 5.0)$. You may observe how the accuracy depends on the interval.

```
PROGRAM d4r7(input,output);
(* driver for routine QGAUS *)
(*$I MODFILE.PAS *)
CONST
   x1 = 0.0;
   x2 = 5.0;
   nval = 10;
VAR
   dx,ss,x: real;
   i: integer;
FUNCTION func(x: real): real;
BEGIN
   func := x*exp(-x)
END;
(*$I QGAUS.PAS *)
BEGIN
   dx := (x2-x1)/nval;
   writeln;
   writeln('0.0 to','qgaus':10,'expected':13);
   writeln;
   FOR i := 1 TO nval DO BEGIN
      x := x1+i*dx;
      qgaus(x1,x,ss);
      writeln(x:5:2,ss:12:6,(-(1.0+x)*exp(-x)
         +(1.0+x1)*exp(-x1)):12:6)
   END
END.
```

Sample program d4r8, which drives gauleg, performs the same method of quadrature, and on the same function. However, it chooses its own abscissas and weights for the Gauss-Legendre calculation, and is not restricted to a ten-point formula; it can do an N-point calculation for any N. The N abscissas and weights appropriate to an interval $x = 0.0$ to 1.0 are found by sample program d4r8 for the case $N = 10$. The results you should find are listed below. Next the program applies these values to a quadrature and compares the result to that from a formal integration.

position	weight
.013047	.033336
.067468	.074726
.160295	.109543
.283302	.134633
.425563	.147762
.574437	.147762
.716698	.134633
.839705	.109543
.932532	.074726
.986953	.033336

```
PROGRAM d4r8(input,output);
(* driver for routine GAULEG *)
(*$I MODFILE.PAS *)
CONST
```

```
   npoint = 10;
   x1 = 0.0;
   x2 = 1.0;
   x3 = 10.0;
   nval = 10;
TYPE
   DoubleArrayNP = ARRAY [1..npoint] OF double;
VAR
   i,j: integer;
   xx: double;
   x,w: DoubleArrayNP;
FUNCTION func(x: double): double;
BEGIN
   func := x*exp(-x)
END;
(*$I GAULEG.PAS *)
BEGIN
   gauleg(x1,x2,x,w,npoint);
   writeln;
   writeln('#':2,'x[i]':11,'w[i]':12);
   FOR i := 1 TO npoint DO
      writeln(i:2,x[i]:12:6,w[i]:12:6);
(* demonstrate the use of gauleg for an integral *)
   gauleg(x1,x3,x,w,npoint);
   xx := 0.0;
   FOR i := 1 TO npoint DO
      xx := xx+w[i]*func(x[i]);
   writeln;
   writeln('Integral from GAULEG: ',xx:12:6);
   writeln('Actual value: ',((1.0+x1)*exp(-x1)-(1.0+x3)*exp(-x3)):12:6)
END.
```

Chapter 4 of *Numerical Recipes* ends with a short discussion of multidimensional integration, exemplified by routine quad3d which does a 3-dimensional integration by repeated 1-dimensional integration. Recall that the Pascal version of this algorithm is simpler than the FORTRAN version because recursion can be used. Sample program d4r9 applies the method to the integration of func $= x^2 + y^2 + z^2$ over a spherical volume with a radius xmax which is taken successively as $0.1, 0.2, ..., 1.0$. The integral is done in Cartesian rather than spherical coordinates, but the result is compared to that found easily in spherical coordinates, $4\pi(\text{xmax})^5/5$. Procedure func generates the function. Procedures y1 and y2 supply the two limits of the y-integration for each value of x. Similarly z1 and z2 give the limits of z-integration for given x and y.

```
PROGRAM d4r9(input,output);
(* driver for routine QUAD3D *)
(*$I MODFILE.PAS *)
CONST
   pi = 3.1415926;
   nval = 10;
VAR
   i: integer;
   s,xmax,xmin,xmax5: real;
   Quad3dX,Quad3dY: real;
FUNCTION func(x,y,z: real): real;
BEGIN
   func := sqr(x)+sqr(y)+sqr(z)
```

```
END;
FUNCTION z1(x,y: real): real;
BEGIN
   z1 := -sqrt(sqr(xmax)-sqr(x)-sqr(y))
END;
FUNCTION z2(x,y: real): real;
BEGIN
   z2 := sqrt(sqr(xmax)-sqr(x)-sqr(y))
END;
FUNCTION y1(x: real): real;
BEGIN
   y1 := -sqrt(sqr(xmax)-sqr(x))
END;
FUNCTION y2(x: real): real;
BEGIN
   y2 := sqrt(sqr(xmax)-sqr(x))
END;
(*$I QUAD3D.PAS *)
BEGIN
   writeln('Integral of r^2 over a spherical volume');
   writeln;
   writeln('radius':13,'QUAD3D':9,'Actual':10);
   FOR i := 1 TO nval DO BEGIN
      xmax := 0.1*i;
      xmin := -xmax;
      quad3d(xmin,xmax,s);
      xmax5 := sqr(sqr(xmax))*xmax;
      writeln(xmax:12:2,s:10:4,4.0*pi*(xmax5)/5.0:10:4)
   END
END.
```

Chapter 5: Evaluation of Functions

Chapter 5 of Numerical Recipes treats the approximation and evaluation of functions. The methods, along with a few others, are applied in Chapter 6 to the calculation of a collection of "special" functions. Polynomial or power series expansions are perhaps the most often used approximations and a few tips are given for accelerating the convergence of some series. In the case of alternating series, Euler's transformation is popular, and is implemented in program eulsum. *For general polynomials,* ddpoly *demonstrates the evaluation of both the polynomial and its derivatives from a list of its coefficients. The division of one polynomial into another, giving a quotient and remainder polynomial, is done by* poldiv.*

The approximation of functions by Chebyshev polynomial series is presented as a method of arriving at the approximation of nearly smallest deviation from the true function over a given region for a specified order of approximation. The coefficients for such polynomials are given by chebft *and function approximations are subsequently carried out by* chebev. *To generate the derivative or integral of a function from its Chebyshev coefficients, use* chder *or* chint *respectively. Finally, to convert Chebyshev coefficients into coefficients of a polynomial for the same function (a dangerous procedure about which we offer due warning in the text) use* chebpc *and* pcshft *in succession.*

Chapter 5 also treats several methods for which we supply no programs. These are continued fractions, rational functions, recurrence relations, and the solution of quadratic and cubic equations.

$$\star \quad \star \quad \star \quad \star$$

Procedure eulsum applies Euler's transformation to the summation of an alternating series. It is called successively for each term to be summed. Our sample program d5r1 evaluates the approximation

$$\ln(1+x) = x - \frac{x^2}{2} + \frac{x^3}{3} - \frac{x^4}{4} + \cdots \qquad -1 < x < 1$$

It asks how many terms mval are to be included in the approximation and then makes mval calls to eulsum. Each time, index j increases and term takes the value $(-1)^{j+1}x^j/j$. Both this approximation and the function $\ln(1+x)$ itself are evaluated across the region -1 to 1 for comparison. If mval is set less than 1 or more than 40, the program terminates.

```
PROGRAM d5r1(input,output);
(* driver for routine EULSUM *)
(*$I MODFILE.PAS *)
LABEL 99;
CONST
   nval = 40;
TYPE
   RealArrayNVAL = ARRAY [1..nval] OF real;
VAR
   i,j,mval,EulsumNterm: integer;
   sum,term,x,xpower: real;
   EulsumWksp: RealArrayNVAL;
(*$I EULSUM.PAS *)
BEGIN
(* evaluate ln(1+x) := x-x^2/2+x^3/3-x^4/4 ... FOR -1<x<1 *)
   WHILE true DO BEGIN
      writeln;
      writeln('How many terms in polynomial?');
      writeln('Enter n between 1 and ',nval:2,'. (n := 0 to END)');
      readln(mval);
      writeln;
      IF (mval <= 0) OR (mval > nval) THEN GOTO 99;
      writeln('x':9,'actual':14,'polynomial':14);
      FOR i := -8 TO 8 DO BEGIN
         x := i/10.0;
         sum := 0.0;
         xpower := -1;
         FOR j := 1 TO mval DO BEGIN
            xpower := -x*xpower;
            term := xpower/j;
            eulsum(sum,term,j)
         END;
         writeln(x:12:6,ln(1.0+x):12:6,sum:12:6)
      END
   END;
99:
END.
```

ddpoly evaluates a polynomial and its derivatives, given the coefficients of the polynomial in the form of an input vector. Sample program d5r2 illustrates this for the polynomial:

$$(x-1)^5 = -1 + 5x - 10x^2 + 10x^3 - 5x^4 + x^5$$

(This is a foolish example, of course. No one would knowingly evaluate $(x-1)^5$ by multiplying it out and evaluating terms individually—but it gives us a convenient way to check the result!). Since this is a fifth order polynomial, we set nc, the number of coeffients, to 6, and initialize the array c of coefficients, with c[1] being the constant coefficient and c[6] the highest-order coefficient. There are two loops, one of which evaluates for x values from 0.0 to 2.0, and the other of which stores the value of the function and nc-1 derivatives. d[j,i] keeps the entire array of values for printing. In the second part of the program, the polynomial evaluations are compared with

$$f^{(n-1)}(x) = \frac{5!}{(6-n)!}(x-1.0)^{6-n} \qquad n = 1,\ldots,5$$

```
PROGRAM d5r2(input,output);
(* driver for routine DDPOLY *)
(* polynomial (x-1)**5 *)
(*$I MODFILE.PAS *)
CONST
   nc = 6;
   nd = 5;    (* nd=nc-1 *)
   np = 20;
TYPE
   CharArray15 = PACKED ARRAY [1..15] OF char;
   IntegerArrayNC = ARRAY [1..nc] OF real;
   IntegerArrayND = ARRAY [1..nd] OF real;
   RealArrayNDbyNP = ARRAY [1..nd,1..np] OF real;
   RealArray33 = ARRAY [1..33] OF real;
VAR
   i,j: integer;
   x: real;
   c: IntegerArrayNC;
   pd: IntegerArrayND;
   d: RealArrayNDbyNP;
   a: ARRAY [1..nd] OF CharArray15;
   FactrlNtop: integer;
   FactrlA: RealArray33;
FUNCTION power(x: real; n: integer): real;
BEGIN
   IF n = 0 THEN power := 1.0 ELSE power := x*power(x,n-1)
END;
(*$I GAMMLN.PAS *)
(*$I FACTRL.PAS *)
(*$I DDPOLY.PAS *)
BEGIN
   FactrlNtop := 0;
   FactrlA[1] := 1.0;
   a[1] := 'polynomial:    '; a[2] := 'first deriv:   ';
   a[3] := 'second deriv:  '; a[4] := 'third deriv:   ';
   a[5] := 'fourth deriv:  ';
   c[1] := -1.0; c[2] := 5.0; c[3] := -10.0;
   c[4] := 10.0; c[5] := -5.0; c[6] := 1.0;
   FOR i := 1 TO np DO BEGIN
      x := 0.1*i;
      ddpoly(c,nc,x,pd,nc-1);
      FOR j := 1 TO nc-1 DO d[j,i] := pd[j]
   END;
   FOR i := 1 TO nc-1 DO BEGIN
      writeln(' ':7,a[i]);
      writeln('x':12,'DDPOLY':17,'actual':15);
      FOR j := 1 TO np DO BEGIN
         x := 0.1*j;
         writeln(x:15:6,d[i,j]:15:6,
            (factrl(nc-1)/factrl(nc-i))*power(x-1.0,nc-i):15:6)
      END;
      writeln('press ENTER to continue...');
      readln
   END
END.
```

poldiv divides polynomials. Given the coefficients of a numerator and denominator polynomial, `poldiv` returns the coefficients of a quotient and a remainder polynomial. Sample program d5r3 takes

$$\text{Numerator} = u = -1 + 5x - 10x^2 + 10x^3 - 5x^4 + x^5 = (x-1)^5$$

$$\text{Denominator} = v = 1 + 3x + 3x^2 + x^3 = (x+1)^3$$

for which we expect

$$\text{Quotient} = q = 31 - 8x + x^2$$

$$\text{Remainder} = r = -32 - 80x - 80x^2$$

The program compares these with the output of `poldiv`.

```
PROGRAM d5r3(input,output);
(* driver for routine POLDIV *)
(* (x-1)**5/(x+1)**3 *)
(*$I MODFILE.PAS *)
CONST
   n = 6;
   nv = 4;
TYPE
   RealArrayNP = ARRAY [1..n] OF real;
   RealArrayNV = ARRAY [1..nv] OF real;
VAR
   i: integer;
   u,q,r: RealArrayNP;
   v: RealArrayNV;
(*$I POLDIV.PAS *)
BEGIN
   u[1] := -1.0; u[2] := 5.0; u[3] := -10.0;
   u[4] := 10.0; u[5] := -5.0; u[6] := 1.0;
   v[1] := 1.0; v[2] := 3.0; v[3] := 3.0; v[4] := 1.0;
   poldiv(u,n,v,nv,q,r);
   writeln;
   writeln('x^0':10,'x^1':10,'x^2':10,'x^3':10,'x^4':10,'x^5':10);
   writeln;
   writeln('quotient polynomial coefficients:');
   FOR i := 1 TO 6 DO write(q[i]:10:2);
   writeln;
   writeln('expected quotient coefficients:');
   writeln(31.0:10:2,-8.0:10:2,1.0:10:2,0.0:10:2,0.0:10:2,0.0:10:2);
   writeln;
   writeln('remainder polynomial coefficients:');
   FOR i := 1 TO 4 DO write(r[i]:10:2);
   writeln;
   writeln('expected remainder coefficients:');
   writeln(-32.0:10:2,-80.0:10:2,-80.0:10:2,0.0:10:2)
END.
```

The remaining six programs all deal with Chebyshev polynomials. `chebft` evaluates the coefficients for a Chebyshev polynomial approximation of a function on a specified interval and for a maximum degree N of polynomial. Demonstration program d5r4 uses the function `func` $= x^2(x^2 - 2)\sin x$ on the interval $(-\pi/2, \pi/2)$ with the maximum degree of nval=40. Notice that `chebft` is called with this maximum degree specified, even though subsequent evaluations may truncate the Chebyshev series at much lower terms. After we choose the number **mval** of terms in the evaluation, the Chebyshev polynomial is evaluated term by term, for x values between -0.8π and 0.8π, and the result **f** is compared to the actual function value.

```pascal
PROGRAM d5r4(input,output);
(* driver for routine CHEBFT *)
(*$I MODFILE.PAS *)
LABEL 99;
CONST
   nval = 40;
   pio2 = 1.5707963;
   eqs = 1.0e-6;
TYPE
   RealArrayNP = ARRAY [1..nval] OF real;
VAR
   a,b,dum,f: real;
   t0,t1,term,x,y: real;
   i,j,mval: integer;
   c: RealArrayNP;
FUNCTION func(x: real): real;
BEGIN
   func := sqr(x)*(sqr(x)-2.0)*sin(x)
END;
(*$I CHEBFT.PAS *)
BEGIN
   a := -pio2;
   b := pio2;
   chebft(a,b,c,nval);
(* test result *)
   WHILE true DO BEGIN
      writeln;
      writeln('How many terms in Chebyshev evaluation?');
      write('Enter n between 6 and ',nval:2,'. (n := 0 to end).  ');
      readln(mval);
      IF (mval <= 0) OR (mval > nval) THEN GOTO 99;
      writeln;
      writeln('x':9,'actual':14,'chebyshev fit':16);
      FOR i := -8 TO 8 DO BEGIN
         x := i*pio2/10.0;
         y := (x-0.5*(b+a))/(0.5*(b-a));
(* evaluate chebyshev polynomial without using routine chebev *);
         t0 := 1.0;
         t1 := y;
         f := c[2]*t1+c[1]*0.5;
         FOR j := 3 TO mval DO BEGIN
            dum := t1;
            t1 := 2.0*y*t1-t0;
            t0 := dum;
            term := c[j]*t1;
            f := f+term
         END;
         writeln(x:12:6,func(x):12:6,f:12:6)
      END
   END;
99:
END.
```

chebev is the Chebyshev polynomial evaluator and the next sample program d5r5 uses it for the same problem just discussed. In fact, the program is identical except that it replaces the internal polynomial summation with chebev, which applies Clenshaw's recurrence to find the polynomial values.

```
PROGRAM d5r5(input,output);
(* driver for routine CHEBEV *)
(*$I MODFILE.PAS *)
LABEL 99;
CONST
   nval = 40;
   pio2 = 1.5707963;
TYPE
   RealArrayNP = ARRAY [1..nval] OF real;
VAR
   a,b,x: real;
   i,mval: integer;
   c: RealArrayNP;
FUNCTION func(x: real): real;
BEGIN
   func := sqr(x)*(sqr(x)-2.0)*sin(x)
END;
(*$I CHEBFT.PAS *)
(*$I CHEBEV.PAS *)
BEGIN
   a := -pio2;
   b := pio2;
   chebft(a,b,c,nval);
(* test chebyshev evaluation routine *)
   WHILE true DO BEGIN
      writeln;
      writeln('How many terms in Chebyshev evaluation?');
      write('Enter n between 6 and ',nval:2,'. (n := 0 to end).   ');
      readln(mval);
      IF (mval <= 0) OR (mval > nval) THEN GOTO 99;
      writeln;
      writeln('x':9,'actual':14,'chebyshev fit':16);
      FOR i := -8 TO 8 DO BEGIN
         x := i*pio2/10.0;
         writeln(x:12:6,func(x):12:6,chebev(a,b,c,mval,x):12:6)
      END
   END;
99:
END.
```

By the same token, the tests for chint and chder needn't be much different. chint determines Chebyshev coefficients for the integral of the function, and chder for the derivative of the function, given the Chebyshev coefficients for the function itself (from chebft) and the interval (A, B) of evaluation. When applied to the function above, the true integral is

$$\text{fint} = 4x(x^2 - 7)\sin x - (x^4 - 14x^2 + 28)\cos x$$

and the true derivative is

$$\text{fder} = 4x(x^2 - 1)\sin x + x^2(x^2 - 2)\cos x$$

The code in sample programs d5r6 and d5r7 compares the true and Chebyshev-derived integral and derivative values for a range of x in the interval of evaluation. Since chint and chder return Chebyshev coefficients, and not the integral and derivative values themselves, calls to chebev are required for the comparison.

```pascal
PROGRAM d5r6(input,output);
(* driver for routine CHINT *)
(*$I MODFILE.PAS *)
LABEL 99;
CONST
   nval = 40;
   pio2 = 1.5707963;
TYPE
   RealArrayNP = ARRAY [1..nval] OF real;
VAR
   a,b,x: real;
   i,mval: integer;
   c,cint: RealArrayNP;
FUNCTION func(x: real): real;
BEGIN
   func := sqr(x)*(sqr(x)-2.0)*sin(x)
END;
FUNCTION fint(x: real): real;
BEGIN
   fint := 4.0*x*(sqr(x)-7.0)*sin(x)
      -(sqr(sqr(x))-14.0*sqr(x)+28.0)*cos(x)
END;
(*$I CHEBFT.PAS *)
(*$I CHEBEV.PAS *)
(*$I CHINT.PAS *)
BEGIN
   a := -pio2;
   b := pio2;
   chebft(a,b,c,nval);
(* test integral *)
   WHILE true DO BEGIN
      writeln;
      writeln('How many terms in Chebyshev evaluation?');
      write('Enter n between 6 and ',nval:2,'. (n := 0 to end).   ');
      readln(mval);
      IF (mval <= 0) OR (mval > nval) THEN GOTO 99;
      chint(a,b,c,cint,mval);
      writeln;
      writeln('x':9,'actual':14,'Cheby. integ.':16);
      FOR i := -8 TO 8 DO BEGIN
         x := i*pio2/10.0;
         writeln(x:12:6,fint(x)-fint(-pio2):12:6,
               chebev(a,b,cint,mval,x):12:6)
      END
   END;
99:
END.


PROGRAM d5r7(input,output);
(* driver for routine CHDER *)
(*$I MODFILE.PAS *)
LABEL 99;
```

```
CONST
   nval = 40;
   pio2 = 1.5707963;
TYPE
   RealArrayNP = ARRAY [1..nval] OF real;
VAR
   a,b,x: real;
   i,mval: integer;
   c,cder: RealArrayNP;
FUNCTION func(x: real): real;
BEGIN
   func := sqr(x)*(sqr(x)-2.0)*sin(x)
END;
FUNCTION fder(x: real): real;
BEGIN
   fder := 4.0*x*(sqr(x)-1.0)*sin(x)+sqr(x)*(sqr(x)-2.0)*cos(x)
END;
(*$I CHEBEV.PAS *)
(*$I CHEBFT.PAS *)
(*$I CHDER.PAS *)
BEGIN
   a := -pio2;
   b := pio2;
   chebft(a,b,c,nval);
(* test derivative *)
   WHILE true DO BEGIN
      writeln;
      writeln('How many terms in Chebyshev evaluation?');
      write('Enter n between 6 and ',nval:2,'. (n := 0 to end).  ');
      readln(mval);
      IF (mval <= 0) OR (mval > nval) THEN GOTO 99;
      chder(a,b,c,cder,mval);
      writeln;
      writeln('x':9,'actual':14,'Cheby. deriv.':16);
      FOR i := -8 TO 8 DO BEGIN
         x := i*pio2/10.0;
         writeln(x:12:6,fder(x):12:6,chebev(a,b,cder,mval,x):12:6)
      END
   END;
99:
END.
```

The final two programs of this chapter turn the coefficients of a Chebyshev approximation into those of a polynomial approximation in the variable

$$y = \frac{x - \frac{1}{2}(B + A)}{\frac{1}{2}(B - A)}$$

(routine chebpc), or of a polynomial approximation in x itself (routine chebpc followed by pcshft). These procedures are discouraged for reasons discussed in *Numerical Recipes*, but should they serve some special purpose for you, we have at least warned that you will be sacrificing accuracy, particularly for polynomials above order 7 or 8. Sample program d5r8 calls chebft and chebpc to find polynomial coefficients in y for a truncated series. For a set of x values between $-\pi$ and π it calculates y and then the terms of the y-polynomial, which are summed in variable poly. Finally, poly is compared to the true function value. (The function func is the same used before.)

```pascal
PROGRAM d5r8(input,output);
(* driver for routine CHEBPC *)
(*$I MODFILE.PAS *)
LABEL 99;
CONST
   nval = 40;
   pio2 = 1.5707963;
TYPE
   RealArrayNP = ARRAY [1..nval] OF real;
VAR
   a,b,poly,x,y: real;
   i,j,mval: integer;
   c,d: RealArrayNP;
FUNCTION func(x: real): real;
BEGIN
   func := sqr(x)*(sqr(x)-2.0)*sin(x)
END;
(*$I CHEBFT.PAS *)
(*$I CHEBPC.PAS *)
BEGIN
   a := -pio2;
   b := pio2;
   chebft(a,b,c,nval);
   WHILE true DO BEGIN
      writeln;
      writeln('How many terms in Chebyshev evaluation?');
      write('Enter n between 6 and ',nval:2,'. (n := 0 to end).   ');
      readln(mval);
      IF (mval <= 0) OR (mval > nval) THEN GOTO 99;
      chebpc(c,d,mval);
(* test polynomial *)
      writeln;
      writeln('x':9,'actual':14,'polynomial':14);
      FOR i := -8 TO 8 DO BEGIN
         x := i*pio2/10.0;
         y := (x-(0.5*(b+a)))/(0.5*(b-a));
         poly := d[mval];
         FOR j := mval-1 DOWNTO 1 DO
            poly := poly*y+d[j];
         writeln(x:12:6,func(x):12:6,poly:12:6)
      END
   END;
99:
END.
```

pcshft shifts the polynomial to be one in variable x. Sample program d5r9 is like the previous program except that it follows the call to chebpc with a call to pcshft.

```pascal
PROGRAM d5r9(input,output);
(* driver for routine PCSHFT *)
(*$I MODFILE.PAS *)
LABEL 99;
CONST
   nval = 40;
```

```
      pio2 = 1.5707963;
TYPE
   RealArrayNP = ARRAY [1..nval] OF real;
VAR
   a,b,poly,x: real;
   i,j,mval: integer;
   c,d: RealArrayNP;
FUNCTION func(x: real): real;
BEGIN
   func := sqr(x)*(sqr(x)-2.0)*sin(x)
END;
(*$I CHEBFT.PAS *)
(*$I CHEBPC.PAS *)
(*$I PCSHFT.PAS *)
BEGIN
   a := -pio2;
   b := pio2;
   chebft(a,b,c,nval);
   WHILE true DO BEGIN
      writeln;
      writeln('How many terms in Chebyshev evaluation?');
      write('Enter n between 6 and ',nval:2,'. (n := 0 to end).   ');
      readln(mval);
      IF (mval <= 0) OR (mval > nval) THEN GOTO 99;
      chebpc(c,d,mval);
      pcshft(a,b,d,mval);
(* test shifted polynomial *)
      writeln;
      writeln('x':9,'actual':14,'polynomial':14);
      FOR i := -8 TO 8 DO BEGIN
         x := i*pio2/10.0;
         poly := d[mval];
         FOR j := mval-1 DOWNTO 1 DO
            poly := poly*x+d[j];
         writeln(x:12:6,func(x):12:6,poly:12:6)
      END
   END;
99:
END.
```

Chapter 6: Special Functions

This chapter on special functions provides illustrations of techniques developed in Chapter 5. At the same time, it offers routines for calculating many of the functions that arise frequently in analytical work, but which are not so common to be included, for example, as a single keystroke on your pocket calculator. In terms of demonstration programs, they represent a simple bunch. The test routines are all virtually identical, all making reference to a single file of function values called `fncval.dat` which is listed in the Appendix at the end of this chapter. In this file are accurate values for the individual functions for a variety of values for each argument. We have aimed to "stress" the routines a bit by throwing in some extreme values for the arguments.

Many of the function values came from Abramowitz and Stegun's Handbook of Mathematical Functions. Some others, however, came from our library of dusty volumes from past masters. There is an implicit danger in a comparison test like this—namely, that our source has used the same algorithms as ours to construct the tables. In that case, we test only our mutual competence at computing, not the correctness of the result. Nevertheless, there is some assurance in knowing that the values we calculate are the ones that have been used and scrutinized for many years. Moreover, the expressions for the functions themselves can be worked out in certain special or limiting cases without computer aid, and in these instances the results have proven correct.

★   ★   ★   ★

With few exceptions, the routines that follow work in this fashion:

1. Open file `fncval.dat`.

2. Find the appropriate data table according to its title.

3. Read the argument list for each table entry and pass them to the routine to be tested.

4. Print the arguments along with the expected and actual results.

For the routines in this list, therefore, we forego any further comment, but simply identify them by the special function that they evaluate.

Natural logarithm of the gamma function for positive arguments:

```
PROGRAM d6r1(input,output,dfile);
(* driver for routine GAMMLN *)
(*$I MODFILE.PAS *)
```

```
CONST
   pi = 3.1415926;
TYPE
   StrArray14 = string[14];
VAR
   i,nval: integer;
   actual,calc,x,y: real;
   txt: StrArray14;
   dfile: text;
(*$I GAMMLN.PAS *)
BEGIN
   NROpen(dfile,'fncval.dat');
   REPEAT readln(dfile,txt) UNTIL txt = 'Gamma Function';
   readln(dfile,nval);
   writeln('gamma function:');
   writeln('x':10,'actual':21,'from gammln(x)':22);
   FOR i := 1 TO nval DO BEGIN
      readln(dfile,x,actual);
      IF x > 0.0 THEN BEGIN
         IF x >= 1.0 THEN
            calc := exp(gammln(x))
         ELSE
            calc := exp(gammln(x+1.0))/x
      END ELSE BEGIN
         y := 1.0-x;
         calc := pi*exp(-gammln(y))/sin(pi*y)
      END;
      writeln(x:12:2,'          ',actual:13,'          ',calc:13)
   END;
   close(dfile)
END.
```

Factorial function $N!$:

```
PROGRAM d6r2(input,output,dfile);
(* driver for routine FACTRL *)
(*$I MODFILE.PAS *)
TYPE
   StrArray11 = string[11];
   RealArray33 = ARRAY [1..33] OF real;
VAR
   actual: real;
   FactrlNtop,i,n,nval: integer;
   txt: StrArray11;
   dfile: text;
   FactrlA: RealArray33;
(*$I GAMMLN.PAS *)
(*$I FACTRL.PAS *)
BEGIN
   FactrlNtop := 0;   (* initialize FACTRL *)
   FactrlA[1] := 1.0;
   NROpen(dfile,'fncval.dat');
   REPEAT readln(dfile,txt) UNTIL txt = 'N-factorial';
   readln(dfile,nval);
   writeln(txt);
   writeln('n':6,'actual':19,'factrl(n)':21);
   FOR i := 1 TO nval DO BEGIN
```

```
      readln(dfile,n,actual);
      IF actual < 1.0e10 THEN
          writeln(n:6,actual:20:0,factrl(n):20:0)
      ELSE
          writeln(n:6,' ':7,actual:13,' ':7,factrl(n):13)
   END;
   close(dfile)
END.
```

Binomial coefficients:

```
PROGRAM d6r3(input,output,dfile);
(* driver for routine BICO *)
(*$I MODFILE.PAS *)
TYPE
   StrArray21 = string[21];
   RealArray100 = ARRAY [1..100] OF real;
VAR
   binco: real;
   i,k,n,nval: integer;
   txt: StrArray21;
   dfile: text;
   FactlnA: RealArray100;
(*$I GAMMLN.PAS *)
(*$I FACTLN.PAS *)
(*$I BICO.PAS *)
BEGIN
   FOR i := 1 TO 100 DO FactlnA[i] := -1.0; (* initialize FACTLN *)
   NROpen(dfile,'fncval.dat');
   REPEAT readln(dfile,txt) UNTIL txt = 'Binomial Coefficients';
   readln(dfile,nval);
   writeln(txt);
   writeln('n':6,'k':6,'actual':12,'bico(n,k)':12);
   FOR i := 1 TO nval DO BEGIN
      readln(dfile,n,k,binco);
      writeln(n:6,k:6,binco:12:0,bico(n,k):12:0)
   END;
   close(dfile)
END.
```

Natural logarithm of *N*!:

```
PROGRAM d6r4(input,output,dfile);
(* driver for routine FACTLN *)
(*$I MODFILE.PAS *)
TYPE
   StrArray11 = string[11];
   RealArray100 = ARRAY [1..100] OF real;
VAR
   i,n,nval: integer;
   val: real;
   txt: StrArray11;
   dfile: text;
   FactlnA: RealArray100;
(*$I GAMMLN.PAS *)
(*$I FACTLN.PAS *)
BEGIN
   FOR i := 1 TO 100 DO FactlnA[i] := -1.0;  (* initialize FACTLN *)
```

```
    NROpen(dfile,'fncval.dat');
    REPEAT readln(dfile,txt) UNTIL txt = 'N-factorial';
    readln(dfile,nval);
    writeln('log of n-factorial');
    writeln('n':6,'actual':19,'factln(n)':21);
    FOR i := 1 TO nval DO BEGIN
        readln(dfile,n,val);
        writeln(n:6,ln(val):20:7,factln(n):20:7)
    END;
    close(dfile)
END.
```

Beta function:

```
PROGRAM d6r5(input,output,dfile);
(* driver for routine BETA *)
(*$I MODFILE.PAS *)
TYPE
    StrArray13 = string[13];
VAR
    i,nval: integer;
    val,w,z: real;
    txt: StrArray13;
    dfile: text;
(*$I GAMMLN.PAS *)
(*$I BETA.PAS *)
BEGIN
    NROpen(dfile,'fncval.dat');
    REPEAT readln(dfile,txt) UNTIL txt = 'Beta Function';
    readln(dfile,nval);
    writeln(txt);
    writeln('w':5,'z':6,'actual':16,'beta(w,z)':20);
    FOR i := 1 TO nval DO BEGIN
        readln(dfile,w,z,val);
        writeln(w:6:2,z:6:2,' ':5,val:13,' ':5,beta(w,z):13)
    END;
    close(dfile)
 END.
```

Incomplete gamma function $P(a, x)$:

```
PROGRAM d6r6(input,output,dfile);
(* driver for routine GAMMP *)
(*$I MODFILE.PAS *)
TYPE
    StrArray25 = string[25];
VAR
    a,val,x: real;
    i,nval: integer;
    txt: StrArray25;
    dfile: text;
(*$I GAMMLN.PAS *)
(*$I GCF.PAS *)
(*$I GSER.PAS *)
(*$I GAMMP.PAS *)
BEGIN
    NROpen(dfile,'fncval.dat');
    REPEAT readln(dfile,txt) UNTIL txt = 'Incomplete Gamma Function';
```

```
    readln(dfile,nval);
    writeln(txt);
    writeln('a':4,'x':11,'actual':14,'gammp(a,x)':14);
    FOR i := 1 TO nval DO BEGIN
        readln(dfile,a,x,val);
        writeln(a:6:2,x:12:6,val:12:6,gammp(a,x):12:6)
    END;
    close(dfile)
END.
```

Incomplete gamma function $Q(a, x) = 1 - P(a, x)$:

```
PROGRAM d6r7(input,output,dfile);
(* driver for routine GAMMQ *)
(*$I MODFILE.PAS *)
TYPE
    StrArray25 = string[25];
VAR
    a,val,x: real;
    i,nval: integer;
    txt: StrArray25;
    dfile: text;
(*$I GAMMLN.PAS *)
(*$I GSER.PAS *)
(*$I GCF.PAS *)
(*$I GAMMQ.PAS *)
BEGIN
    NROpen(dfile,'fncval.dat');
    REPEAT readln(dfile,txt) UNTIL txt = 'Incomplete Gamma Function';
    readln(dfile,nval);
    writeln(txt);
    writeln('a':4,'x':11,'actual':14,'gammq(a,x)':14);
    FOR i := 1 TO nval DO BEGIN
        readln(dfile,a,x,val);
        writeln(a:6:2,x:12:6,(1.0-val):12:6,gammq(a,x):12:6)
    END;
    close(dfile)
END.
```

Incomplete gamma function $P(a, x)$ evaluated from series representation:

```
PROGRAM d6r8(input,output,dfile);
(* driver for routine GSER *)
(*$I MODFILE.PAS *)
TYPE
    StrArray25 = string[25];
VAR
    a,gamser,gln,val,x: real;
    i,nval: integer;
    txt: StrArray25;
    dfile: text;
(*$I GAMMLN.PAS *)
(*$I GSER.PAS *)
BEGIN
    NROpen(dfile,'fncval.dat');
    REPEAT readln(dfile,txt) UNTIL txt = 'Incomplete Gamma Function';
    readln(dfile,nval);
    writeln(txt);
```

```
    writeln('a':4,'x':11,'actual':14,'gser(a,x)':14,
        'gammln(a)':12,'gln':8);
    FOR i := 1 TO nval DO BEGIN
        readln(dfile,a,x,val);
        gser(a,x,gamser,gln);
        writeln(a:6:2,x:12:6,val:12:6,gamser:12:6,
            gammln(a):12:6,gln:12:6)
    END;
    close(dfile)
END.
```

Incomplete gamma function $Q(a, x)$ evaluated by continued fraction representation:

```
PROGRAM d6r9(input,output,dfile);
(* driver for routine GCF *)
(*$I MODFILE.PAS *)
TYPE
    StrArray25 = string[25];
VAR
    a,gammcf,gln,val,x: real;
    i,nval: integer;
    txt: StrArray25;
    dfile: text;
(*$I GAMMLN.PAS *)
(*$I GCF.PAS *)
BEGIN
    NROpen(dfile,'fncval.dat');
    REPEAT readln(dfile,txt) UNTIL txt = 'Incomplete Gamma Function';
    readln(dfile,nval);
    writeln(txt);
    writeln('a':4,'x':11,'actual':14,'gcf(a,x)':13,
        'gammln(a)':13,'gln':8);
    FOR i := 1 TO nval DO BEGIN
        readln(dfile,a,x,val);
        IF x >= a+1.0 THEN BEGIN
            gcf(a,x,gammcf,gln);
            writeln(a:6:2,x:12:6,(1.0-val):12:6,
                gammcf:12:6,gammln(a):12:6,gln:12:6)
        END
    END;
    close(dfile)
END.
```

Error function:

```
PROGRAM d6r10(input,output,dfile);
(* driver for routine ERF *)
(*$I MODFILE.PAS *)
TYPE
    StrArray14 = string[14];
VAR
    i,nval: integer;
    val,x: real;
    txt: StrArray14;
    dfile: text;
(*$I GAMMLN.PAS *)
(*$I GSER.PAS *)
(*$I GCF.PAS *)
```

```
(*$I GAMMP.PAS *)
(*$I ERF.PAS *)
BEGIN
   NROpen(dfile,'fncval.dat');
   REPEAT readln(dfile,txt) UNTIL txt = 'Error Function';
   readln(dfile,nval);
   writeln(txt);
   writeln('x':5,'actual':12,'erf(x)':12);
   FOR i := 1 TO nval DO BEGIN
      readln(dfile,x,val);
      writeln(x:6:2,val:12:7,erf(x):12:7)
   END;
   close(dfile)
END.
```

Complementary error function:

```
PROGRAM d6r11(input,output,dfile);
(* driver for routine ERFC *)
(*$I MODFILE.PAS *)
TYPE
   StrArray14 = string[14];
VAR
   i,nval: integer;
   val,x: real;
   txt: StrArray14;
   dfile: text;
(*$I GAMMLN.PAS *)
(*$I GCF.PAS *)
(*$I GSER.PAS *)
(*$I GAMMQ.PAS *)
(*$I GAMMP.PAS *)
(*$I ERFC.PAS *)
BEGIN
   NROpen(dfile,'fncval.dat');
   REPEAT readln(dfile,txt) UNTIL txt = 'Error Function';
   readln(dfile,nval);
   writeln('complementary error function');
   writeln('x':5,'actual':12,'erfc(x)':12);
   FOR i := 1 TO nval DO BEGIN
      readln(dfile,x,val);
      val := 1.0-val;
      writeln(x:6:2,val:12:7,erfc(x):12:7)
   END;
   close(dfile);
END.
```

Complementary error function from a Chebyshev fit to a guessed functional form:

```
PROGRAM d6r12(input,output,dfile);
(* driver for routine erfcc *)
(*$I MODFILE.PAS *)
TYPE
   StrArray14 = string[14];
VAR
   i,nval: integer;
   x,val: real;
   txt: StrArray14;
```

```
   dfile: text;
(*$I ERFCC.PAS *)
BEGIN
   NROpen(dfile,'fncval.dat');
   REPEAT readln(dfile,txt) UNTIL txt = 'Error Function';
   readln(dfile,nval);
   writeln('complementary error function');
   writeln('x':5,'actual':12,'erfcc(x)':13);
   FOR i := 1 TO nval DO BEGIN
      readln(dfile,x,val);
      val := 1.0-val;
      writeln(x:6:2,val:12:7,erfcc(x):12:7)
   END;
   close(dfile)
END.
```

Incomplete Beta function:

```
PROGRAM d6r13(input,output,dfile);
(* driver for routines BETAI and BETACF *)
(*$I MODFILE.PAS *)
TYPE
   StrArray24 = string[24];
VAR
   a,b,val,x: real;
   i,nval: integer;
   txt: StrArray24;
   dfile: text;
(*$I GAMMLN.PAS *)
(*$I BETACF.PAS *)
(*$I BETAI.PAS *)
BEGIN
   NROpen(dfile,'fncval.dat');
   REPEAT readln(dfile,txt) UNTIL txt = 'Incomplete Beta Function';
   readln(dfile,nval);
   writeln(txt);
   writeln('a':5,'b':10,'x':12,'actual':14,'betai(x)':13);
   FOR i := 1 TO nval DO BEGIN
      readln(dfile,a,b,x,val);
      writeln(a:6:2,b:12:6,x:12:6,val:12:6,betai(a,b,x):12:6)
   END;
   close(dfile)
END.
```

Bessel function J_0:

```
PROGRAM d6r15(input,output,dfile);
(* driver for routine BESSJ0 *)
(*$I MODFILE.PAS *)
TYPE
   StrArray18 = string[18];
VAR
   i,nval: integer;
   val,x: real;
   txt: StrArray18;
   dfile: text;
(*$I BESSJ0.PAS *)
BEGIN
```

```
   NROpen(dfile,'fncval.dat');
   REPEAT readln(dfile,txt) UNTIL txt = 'Bessel Function J0';
   readln(dfile,nval);
   writeln(txt);
   writeln('x':5,'actual':12,'bessj0(x)':13);
   FOR i := 1 TO nval DO BEGIN
      readln(dfile,x,val);
      writeln(x:6:2,val:12:7,bessj0(x):12:7)
   END;
   close(dfile)
END.
```

Bessel function Y_0:

```
PROGRAM d6r16(input,output,dfile);
(* driver for routine BESSY0 *)
(*$I MODFILE.PAS *)
TYPE
   StrArray18 = string[18];
VAR
   i,nval: integer;
   val,x: real;
   txt: StrArray18;
   dfile: text;
(*$I BESSJ0.PAS *)
(*$I BESSY0.PAS *)
BEGIN
   NROpen(dfile,'fncval.dat');
   REPEAT readln(dfile,txt) UNTIL txt = 'Bessel Function Y0';
   readln(dfile,nval);
   writeln(txt);
   writeln('x':5,'actual':12,'bessy0(x)':13);
   FOR i := 1 TO nval DO BEGIN
      readln(dfile,x,val);
      writeln(x:6:2,val:12:7,bessy0(x):12:7)
   END;
   close(dfile)
END.
```

Bessel function J_1:

```
PROGRAM d6r17(input,output,dfile);
(* driver for routine BESSJ1 *)
(*$I MODFILE.PAS *)
TYPE
   StrArray18 = string[18];
VAR
   i,nval: integer;
   val,x: real;
   txt: StrArray18;
   dfile: text;
(*$I BESSJ1.PAS *)
BEGIN
   NROpen(dfile,'fncval.dat');
   REPEAT readln(dfile,txt) UNTIL txt = 'Bessel Function J1';
   readln(dfile,nval);
   writeln(txt);
   writeln('x':5,'actual':12,'bessj1(x)':13);
```

```
    FOR i := 1 TO nval DO BEGIN
        readln(dfile,x,val);
        writeln(x:6:2,val:12:7,bessj1(x):12:7)
    END;
    close(dfile);
END.
```

Bessel function Y_1:

```
PROGRAM d6r18(input,output,dfile);
(* driver for routine BESSY1 *)
(*$I MODFILE.PAS *)
TYPE
    StrArray18 = string[18];
VAR
    i,nval: integer;
    val,x: real;
    txt: StrArray18;
    dfile: text;
(*$I BESSJ1.PAS *)
(*$I BESSY1.PAS *)
BEGIN
    NROpen(dfile,'fncval.dat');
    REPEAT readln(dfile,txt) UNTIL txt = 'Bessel Function Y1';
    readln(dfile,nval);
    writeln(txt);
    writeln('x':5,'actual':12,'bessy1(x)':13);
    FOR i := 1 TO nval DO BEGIN
        readln(dfile,x,val);
        writeln(x:6:2,val:12:7,bessy1(x):12:7)
    END;
    close(dfile)
END.
```

Bessel function Y_n for $n > 1$:

```
PROGRAM d6r19(input,output,dfile);
(* driver for routine BESSY *)
(*$I MODFILE.PAS *)
TYPE
    StrArray18 = string[18];
VAR
    i,n,nval: integer;
    val,x: real;
    txt: StrArray18;
    dfile: text;
(*$I BESSJ0.PAS *)
(*$I BESSJ1.PAS *)
(*$I BESSY0.PAS *)
(*$I BESSY1.PAS *)
(*$I BESSY.PAS *)
BEGIN
    NROpen(dfile,'fncval.dat');
    REPEAT readln(dfile,txt) UNTIL txt = 'Bessel Function Yn';
    readln(dfile,nval);
    writeln(txt);
    writeln('n':4,'x':7,'actual':14,'bessy(n,x)':18);
    FOR i := 1 TO nval DO BEGIN
```

```
      readln(dfile,n,x,val);
      writeln(n:4,x:8:2,'    ',val:13,'    ',bessy(n,x):13)
   END;
   close(dfile)
END.
```

Bessel function J_n for $n > 1$:

```
PROGRAM d6r20(input,output,dfile);
(* driver for routine BESSJ *)
(*$I MODFILE.PAS *)
TYPE
   StrArray18 = string[18];
VAR
   i,n,nval: integer;
   val,x: real;
   txt: StrArray18;
   dfile: text;
(*$I BESSJ0.PAS *)
(*$I BESSJ1.PAS *)
(*$I BESSJ.PAS *)
BEGIN
   NROpen(dfile,'fncval.dat');
   REPEAT readln(dfile,txt) UNTIL txt = 'Bessel Function Jn';
   readln(dfile,nval);
   writeln(txt);
   writeln('n':4,'x':7,'actual':14,'bessj(n,x)':18);
   FOR i := 1 TO nval DO BEGIN
      readln(dfile,n,x,val);
      writeln(n:4,x:8:2,'    ',val:13,'    ',bessj(n,x):13)
   END;
   close(dfile)
END.
```

Bessel function I_0:

```
PROGRAM d6r21(input,output,dfile);
(* driver for routine BESSI0 *)
(*$I MODFILE.PAS *)
TYPE
   StrArray27 = string[27];
VAR
   i,nval: integer;
   val,x: real;
   txt: StrArray27;
   dfile: text;
(*$I BESSI0.PAS *)
BEGIN
   NROpen(dfile,'fncval.dat');
   REPEAT readln(dfile,txt) UNTIL txt = 'Modified Bessel Function I0';
   readln(dfile,nval);
   writeln(txt);
   writeln;
   writeln('x':5,'actual':16,'bessi0(x)':17);
   FOR i := 1 TO nval DO BEGIN
      readln(dfile,x,val);
      writeln(x:6:2,val:16:7,bessi0(x):16:7)
   END;
```

```
      close(dfile)
END.
```

Bessel function K_0:

```
PROGRAM d6r22(input,output,dfile);
(* driver for routine BESSK0 *)
(*$I MODFILE.PAS *)
TYPE
   StrArray27 = string[27];
VAR
   i,nval: integer;
   val,x: real;
   txt: StrArray27;
   dfile: text;
(*$I BESSI0.PAS *)
(*$I BESSK0.PAS *)
BEGIN
   NROpen(dfile,'fncval.dat');
   REPEAT readln(dfile,txt) UNTIL txt = 'Modified Bessel Function K0';
   readln(dfile,nval);
   writeln(txt);
   writeln;
   writeln('x';5,'actual':16,'bessk0(x)':17);
   FOR i := 1 TO nval DO BEGIN
      readln(dfile,x,val);
      writeln(x:6:2,val:16:7,bessk0(x):16:7)
   END;
   close(dfile)
END.
```

Bessel function I_1:

```
PROGRAM d6r23(input,output,dfile);
(* driver for routine BESSI1 *)
(*$I MODFILE.PAS *)
TYPE
   StrArray27 = string[27];
VAR
   i,nval: integer;
   val,x: real;
   txt: StrArray27;
   dfile: text;
(*$I BESSI1.PAS *)
BEGIN
   NROpen(dfile,'fncval.dat');
   REPEAT readln(dfile,txt) UNTIL txt = 'Modified Bessel Function I1';
   readln(dfile,nval);
   writeln(txt);
   writeln;
   writeln('x':5,'actual':16,'bessi1(x)':17);
   FOR i := 1 TO nval DO BEGIN
      readln(dfile,x,val);
      writeln(x:6:2,val:16:7,bessi1(x):16:7)
   END;
   close(dfile)
END.
```

Bessel function K_1:

```
PROGRAM d6r24(input,output,dfile);
(* driver for routine BESSK1 *)
(*$I MODFILE.PAS *)
TYPE
   StrArray27 = string[27];
VAR
   i,nval: integer;
   val,x: real;
   txt: StrArray27;
   dfile: text;
(*$I BESSI1.PAS *)
(*$I BESSK1.PAS *)
BEGIN
   NROpen(dfile,'fncval.dat');
   REPEAT readln(dfile,txt) UNTIL txt = 'Modified Bessel Function K1';
   readln(dfile,nval);
   writeln(txt);
   writeln;
   writeln('x':5,'actual':16,'bessk1(x)':17);
   FOR i := 1 TO nval DO BEGIN
      readln(dfile,x,val);
      writeln(x:6:2,val:16:7,bessk1(x):16:7)
   END;
   close(dfile)
END.
```

Bessel function K_n for $n > 1$:

```
PROGRAM d6r25(input,output,dfile);
(* driver for routine BESSK *)
(*$I MODFILE.PAS *)
TYPE
   StrArray27 = string[27];
VAR
   n,i,nval: integer;
   val,x: real;
   txt: StrArray27;
   dfile: text;
(*$I BESSI0.PAS *)
(*$I BESSI1.PAS *)
(*$I BESSK0.PAS *)
(*$I BESSK1.PAS *)
(*$I BESSK.PAS *)
BEGIN
   NROpen(dfile,'fncval.dat');
   REPEAT readln(dfile,txt) UNTIL txt = 'Modified Bessel Function Kn';
   readln(dfile,nval);
   writeln(txt);
   writeln;
   writeln('x':5,'actual':14,'bessk(n,x)':18);
   FOR i := 1 TO nval DO BEGIN
      readln(dfile,n,x,val);
      writeln(x:6:2,'   ',val:13,'   ',bessk(n,x):13)
   END;
   close(dfile)
END.
```

Bessel function I_n for $n > 1$:

```pascal
PROGRAM d6r26(input,output,dfile);
(* driver for routine BESSI *)
(*$I MODFILE.PAS *)
TYPE
   StrArray27 = string[27];
VAR
   n,i,nval: integer;
   val,x: real;
   txt: StrArray27;
   dfile: text;
(*$I BESSI0.PAS *)
(*$I BESSI.PAS *)
BEGIN
   NROpen(dfile,'fncval.dat');
   REPEAT readln(dfile,txt) UNTIL txt = 'Modified Bessel Function In';
   readln(dfile,nval);
   writeln(txt);
   writeln;
   writeln('x':5,'actual':14,'bessi(n,x)':18);
   FOR i := 1 TO nval DO BEGIN
      readln(dfile,n,x,val);
      writeln(x:6:2,'    ',val:13,'    ',bessi(n,x):13)
   END;
   close(dfile)
END.
```

Legendre polynomials:

```pascal
PROGRAM d6r27(input,output,dfile);
(* driver for routine PLGNDR *)
(*$I MODFILE.PAS *)
TYPE
   StrArray20 = string[20];
VAR
   fac,val,x: real;
   i,j,m,n,nval: integer;
   txt: StrArray20;
   dfile: text;
(*$I PLGNDR.PAS *)
BEGIN
   NROpen(dfile,'fncval.dat');
   REPEAT readln(dfile,txt) UNTIL txt = 'Legendre Polynomials';
   readln(dfile,nval);
   writeln(txt);
   writeln('n':4,'m':4,'x':10,'actual':19,'plgndr(n,m,x)':20);
   FOR i := 1 TO nval DO BEGIN
      readln(dfile,n,m,x,val);
      fac := 1.0;
      IF m > 0 THEN
         FOR j := n-m+1 TO n+m DO fac := fac*j;
      fac := 2.0*fac/(2.0*n+1.0);
      val := val*sqrt(fac);
      writeln(n:4,m:4,'    ',x:13,'    ',val:13,'    ',plgndr(n,m,x):13)
   END;
   close(dfile)
END.
```

There are three programs that operate in a slightly different fashion. Sample programs d6r28 and d6r29 for routines el2 and cel, which calculate elliptic integrals, do not refer to tables at all. Instead, they make twenty random choices of argument and compare the output of the function evaluation routine for these arguments to the result of actually performing the integration that defines them. The routine qsimp from Chapter 4 of *Numerical Recipes* is used for the integration.

```pascal
PROGRAM d6r28(input,output);
(* driver for routine EL2 *)
(*$I MODFILE.PAS *)
TYPE
   RealArray55 = ARRAY [1..55] OF real;
VAR
   TrapzdIt: integer;
   Ran3Inext,Ran3Inextp: integer;
   Ran3Ma: RealArray55;
   FuncA,FuncB,FuncAkc: real;
   ago,astop,s,x: real;
   i,idum: integer;
FUNCTION func(phi: real): real;
(* Programs using routine FUNC must declare the variables
VAR
   FuncAkc,FuncA,FuncB: real;
in the main routine, and must assign their values. *)
VAR
   tn,tsq: real;
BEGIN
   tn := sin(phi)/cos(phi);
   tsq := tn*tn;
   func := (FuncA+FuncB*tsq)/sqrt((1.0+tsq)*(1.0+FuncAkc*FuncAkc*tsq))
END;
(*$I RAN3.PAS *)
(*$I TRAPZD.PAS *)
(*$I QSIMP.PAS *)
(*$I EL2.PAS *)
BEGIN
   writeln('general elliptic integral of second kind');
   writeln('x':7,'kc':10,'a':10,'b':10,'el2':11,'integral':12);
   idum := -55;
   ago := 0.0;
   FOR i := 1 TO 20 DO BEGIN
      FuncAkc := 5.0*ran3(idum);
      FuncA := 10.0*ran3(idum);
      FuncB := 10.0*ran3(idum);
      x := 10.0*ran3(idum);
      astop := arctan(x);
      qsimp(ago,astop,s);
      writeln(x:10:6,FuncAkc:10:6,FuncA:10:6,FuncB:10:6,
         el2(x,FuncAkc,FuncA,FuncB):10:6,s:10:6)
   END
END.

PROGRAM d6r29(input,output);
(* driver for routine CEL *)
(*$I MODFILE.PAS *)
CONST
   pio2 = 1.5707963;
```

```
     n = 5;
TYPE
     RealArray55 = ARRAY [1..55] OF real;
     RealArrayNP = ARRAY [1..n] OF real;
VAR
     TrapzdIt: integer;
     Ran3Inext,Ran3Inextp: integer;
     Ran3Ma: RealArray55;
     FuncA,FuncB,FuncP,FuncAkc: real;
     ago,astop,s: real;
     i,idum: integer;
FUNCTION func(phi: real): real;
(* Programs using routine FUNC must declare the variables
VAR
     FuncA,FuncB,FuncP,FuncAkc: real;
in the main routine. *)
VAR
     cs,csq,ssq: real;
BEGIN
     cs  := cos(phi);
     csq := cs*cs;
     ssq := 1.0-csq;
     func := (FuncA*csq+FuncB*ssq)/(csq+FuncP*ssq)
         /sqrt(csq+FuncAkc*FuncAkc*ssq)
END;
(*$I RAN3.PAS *)
(*$I TRAPZD.PAS *)
(*$I POLINT.PAS *)
(*$I QROMB.PAS *)
(*$I CEL.PAS *)
BEGIN
     writeln('complete elliptic integral');
     writeln('kc':7,'p':10,'a':10,'b':10,'cel':11,'integral':12);
     idum := -55;
     ago := 0.0;
     astop := pio2;
     FOR i := 1 TO 20 DO BEGIN
        FuncAkc := 0.1+ran3(idum);
        FuncA := 10.0*ran3(idum);
        FuncB := 10.0*ran3(idum);
        FuncP := 0.1+ran3(idum);
        qromb(ago,astop,s);
        writeln(FuncAkc:10:6,FuncP:10:6,FuncA:10:6,FuncB:10:6,
           cel(FuncAkc,FuncP,FuncA,FuncB):10:6,s:10:6)
     END
END.
```

Routine sncndn returns Jacobian elliptic functions. The file fncval.dat contains information only about function sn. However, the values of cn and dn satisfy the relationships

$$\mathrm{sn}^2 + \mathrm{cn}^2 = 1, \qquad k^2\mathrm{sn}^2 + \mathrm{dn}^2 = 1.$$

The program d6r30 works exactly as the others in terms of testing sn, but for verifying cn and dn it lists the values of the left sides of the two equations above. Each of them should have the value 1.0 for all choices of arguments.

```
PROGRAM d6r30(input,output,dfile);
(* driver for routine SNCNDN *)
(*$I MODFILE.PAS *)
TYPE
   StrArray26 = string[26];
VAR
   cn,dn,em,emmc,sn,uu,val: real;
   i,nval: integer;
   txt: StrArray26;
   dfile: text;
(*$I SNCNDN.PAS *)
BEGIN
   NROpen(dfile,'fncval.dat');
   REPEAT readln(dfile,txt) UNTIL txt = 'Jacobian Elliptic Function';
   readln(dfile,nval);
   writeln(txt);
   writeln('mc':4,'u':8,'actual':16,'sn':13,
      'sn^2+cn^2':15,'(mc)*(sn^2)+dn^2':18);
   FOR i := 1 TO nval DO BEGIN
      readln(dfile,em,uu,val);
      emmc := 1.0-em;
      sncndn(uu,emmc,sn,cn,dn);
      writeln(emmc:5:2,uu:8:2,val:15:5,sn:15:5,
         (sn*sn+cn*cn):12:5,(em*sn*sn+dn*dn):14:5)
   END;
   close(dfile)
END.
```

Appendix

File fncval.dat:

```
Values of Special Functions in format x,F(x) or x,y,F(x,y)
Gamma Function
17 Values
   1.0    1.000000
   1.2    0.918169
   1.4    0.887264
   1.6    0.893515
   1.8    0.931384
   2.0    1.000000
   0.2    4.590845
   0.4    2.218160
   0.6    1.489192
   0.8    1.164230
  -0.2    5.2005665E01
  -0.4    4.617091E01
  -0.6    4.0128959E01
  -0.8    3.4231564E01
  10.0    3.6288000E05
  20.0    1.2164510E17
  30.0    8.8417620E30
N-factorial
18 Values
   1      1
   2      2
   3      6
```

```
4       24
5       120
6       720
7       5040
8       40320
9       362880
10      3628800
11      39916800
12      479001600
13      6227020800
14      87178291200
15      1.3076755E12
20      2.4329042E18
25      1.5511222E25
30      2.6525281E32
```

Binomial Coefficients
20 Values

```
1       0       1
6       1       6
6       3       20
6       5       6
15      1       15
15      3       455
15      5       3003
15      7       6435
15      9       5005
15      11      1365
15      13      105
25      1       25
25      3       2300
25      5       53130
25      7       480700
25      9       2042975
25      11      4457400
25      13      5200300
25      15      3268760
25      17      1081575
```

Beta Function
15 Values

```
1.0     1.0     1.000000
0.2     1.0     5.000000
1.0     0.2     5.000000
0.4     1.0     2.500000
1.0     0.4     2.500000
0.6     1.0     1.666667
0.8     1.0     1.250000
6.0     6.0     3.607504E-04
6.0     5.0     7.936508E-04
6.0     4.0     1.984127E-03
6.0     3.0     5.952381E-03
6.0     2.0     0.238095E-01
7.0     7.0     8.325008E-05
5.0     5.0     1.587302E-03
4.0     4.0     7.142857E-03
3.0     3.0     0.333333E-01
2.0     2.0     1.666667E-01
```

Incomplete Gamma Function

20 Values

0.1	3.1622777E-02	0.7420263
0.1	3.1622777E-01	0.9119753
0.1	1.5811388	0.9898955
0.5	7.0710678E-02	0.2931279
0.5	7.0710678E-01	0.7656418
0.5	3.5355339	0.9921661
1.0	0.1000000	0.0951626
1.0	1.0000000	0.6321206
1.0	5.0000000	0.9932621
1.1	1.0488088E-01	0.0757471
1.1	1.0488088	0.6076457
1.1	5.2440442	0.9933425
2.0	1.4142136E-01	0.0091054
2.0	1.4142136	0.4130643
2.0	7.0710678	0.9931450
6.0	2.4494897	0.0387318
6.0	12.247449	0.9825937
11.0	16.583124	0.9404267
26.0	25.495098	0.4863866
41.0	44.821870	0.7359709

Error Function

20 Values

0.0	0.000000
0.1	0.1124629
0.2	0.2227026
0.3	0.3286268
0.4	0.4283924
0.5	0.5204999
0.6	0.6038561
0.7	0.6778012
0.8	0.7421010
0.9	0.7969082
1.0	0.8427008
1.1	0.8802051
1.2	0.9103140
1.3	0.9340079
1.4	0.9522851
1.5	0.9661051
1.6	0.9763484
1.7	0.9837905
1.8	0.9890905
1.9	0.9927904

Incomplete Beta Function

20 Values

0.5	0.5	0.01	0.0637686
0.5	0.5	0.10	0.2048328
0.5	0.5	1.00	1.0000000
1.0	0.5	0.01	0.0050126
1.0	0.5	0.10	0.0513167
1.0	0.5	1.00	1.0000000
1.0	1.0	0.5	0.5000000
5.0	5.0	0.5	0.5000000
10.0	0.5	0.9	0.1516409
10.0	5.0	0.5	0.0897827
10.0	5.0	1.0	1.0000000
10.0	10.0	0.5	0.5000000

```
20.0    5.0     0.8     0.4598773
20.0    10.0    0.6     0.2146816
20.0    10.0    0.8     0.9507365
20.0    20.0    0.5     0.5000000
20.0    20.0    0.6     0.8979414
30.0    10.0    0.7     0.2241297
30.0    10.0    0.8     0.7586405
40.0    20.0    0.7     0.7001783
```

Bessel Function J0
20 Values
```
-5.0    -0.1775968
-4.0    -0.3971498
-3.0    -0.2600520
-2.0     0.2238908
-1.0     0.7651976
 0.0     1.0000000
 1.0     0.7651977
 2.0     0.2238908
 3.0    -0.2600520
 4.0    -0.3971498
 5.0    -0.1775968
 6.0     0.1506453
 7.0     0.3000793
 8.0     0.1716508
 9.0    -0.0903336
10.0    -0.2459358
11.0    -0.1711903
12.0     0.0476893
13.0     0.2069261
14.0     0.1710735
15.0    -0.0142245
```

Bessel Function Y0
15 Values
```
 0.1    -1.5342387
 1.0     0.0882570
 2.0     0.51037567
 3.0     0.37685001
 4.0    -0.0169407
 5.0    -0.3085176
 6.0    -0.2881947
 7.0    -0.0259497
 8.0     0.2235215
 9.0     0.2499367
10.0     0.0556712
11.0    -0.1688473
12.0    -0.2252373
13.0    -0.0782079
14.0     0.1271926
15.0     0.2054743
```

Bessel Function J1
20 Values
```
-5.0     0.3275791
-4.0     0.0660433
-3.0    -0.3390590
-2.0    -0.5767248
-1.0    -0.4400506
 0.0     0.0000000
```

```
 1.0     0.4400506
 2.0     0.5767248
 3.0     0.3390590
 4.0    -0.0660433
 5.0    -0.3275791
 6.0    -0.2766839
 7.0    -0.0046828
 8.0     0.2346364
 9.0     0.2453118
10.0     0.0434728
11.0    -0.1767853
12.0    -0.2234471
13.0    -0.0703181
14.0     0.1333752
15.0     0.2051040
Bessel Function Y1
15 Values
 0.1    -6.4589511
 1.0    -0.7812128
 2.0    -0.1070324
 3.0     0.3246744
 4.0     0.3979257
 5.0     0.1478631
 6.0    -0.1750103
 7.0    -0.3026672
 8.0    -0.1580605
 9.0     0.1043146
10.0     0.2490154
11.0     0.1637055
12.0    -0.0570992
13.0    -0.2100814
14.0    -0.1666448
15.0     0.0210736
Bessel Function Jn, n>=2
20 Values
 2      1.0      1.149034849E-01
 2      2.0      3.528340286E-01
 2      5.0      4.656511628E-02
 2     10.0      2.546303137E-01
 2     50.0     -5.971280079E-02
 5      1.0      2.497577302E-04
 5      2.0      7.039629756E-03
 5      5.0      2.611405461E-01
 5     10.0     -2.340615282E-01
 5     50.0     -8.140024770E-02
10      1.0      2.630615124E-10
10      2.0      2.515386283E-07
10      5.0      1.467802647E-03
10     10.0      2.074861066E-01
10     50.0     -1.138478491E-01
20      1.0      3.873503009E-25
20      2.0      3.918972805E-19
20      5.0      2.770330052E-11
20     10.0      1.151336925E-05
20     50.0     -1.167043528E-01
Bessel Function Yn, n>=2
20 Values
```

```
2      1.0     -1.650682607
2      2.0     -6.174081042E-01
2      5.0      3.676628826E-01
2     10.0     -5.868082460E-03
2     50.0      9.579316873E-02
5      1.0     -2.604058666E02
5      2.0     -9.935989128
5      5.0     -4.536948225E-01
5     10.0      1.354030477E-01
5     50.0     -7.854841391E-02
10     1.0     -1.216180143E08
10     2.0     -1.291845422E05
10     5.0     -2.512911010E01
10    10.0     -3.598141522E-01
10    50.0      5.723897182E-03
20     1.0     -4.113970315E22
20     2.0     -4.081651389E16
20     5.0     -5.933965297E08
20    10.0     -1.597483848E03
20    50.0      1.644263395E-02
```

Modified Bessel Function I0
20 Values

```
0.0     1.0000000
0.2     1.0100250
0.4     1.0404018
0.6     1.0920453
0.8     1.1665149
1.0     1.2660658
1.2     1.3937256
1.4     1.5533951
1.6     1.7499807
1.8     1.9895593
2.0     2.2795852
2.5     3.2898391
3.0     4.8807925
3.5     7.3782035
4.0    11.301922
4.5    17.481172
5.0    27.239871
6.0    67.234406
8.0   427.56411
10.0  2815.7167
```

Modified Bessel Function K0
20 Values

```
0.1     2.4270690
0.2     1.7527038
0.4     1.1145291
0.6     0.77752208
0.8     0.56534710
1.0     0.42102445
1.2     0.31850821
1.4     0.24365506
1.6     0.18795475
1.8     0.14593140
2.0     0.11389387
2.5     6.2347553E-02
3.0     3.4739500E-02
```

```
3.5     1.9598897E-02
4.0     1.1159676E-02
4.5     6.3998572E-03
5.0     3.6910983E-03
6.0     1.2439943E-03
8.0     1.4647071E-04
10.0    1.7780062E-05
```

Modified Bessel Function I1
20 Values

```
0.0     0.00000000
0.2     0.10050083
0.4     0.20402675
0.6     0.31370403
0.8     0.43286480
1.0     0.56515912
1.2     0.71467794
1.4     0.88609197
1.6     1.0848107
1.8     1.3171674
2.0     1.5906369
2.5     2.5167163
3.0     3.9533700
3.5     6.2058350
4.0     9.7594652
4.5     15.389221
5.0     24.335643
6.0     61.341937
8.0     399.87313
10.0    2670.9883
```

Modified Bessel Function K1
20 Values

```
0.1     9.8538451
0.2     4.7759725
0.4     2.1843544
0.6     1.3028349
0.8     0.86178163
1.0     0.60190724
1.2     0.43459241
1.4     0.32083589
1.6     0.24063392
1.8     0.18262309
2.0     0.13986588
2.5     7.3890816E-02
3.0     4.0156431E-02
3.5     2.2239393E-02
4.0     1.2483499E-02
4.5     7.0780949E-03
5.0     4.0446134E-03
6.0     1.3439197E-03
8.0     1.5536921E-04
10.0    1.8648773E-05
```

Modified Bessel Function Kn, n>=2
28 Values

```
2       0.2     49.512430
2       1.0     1.6248389
2       2.0     2.5375975E-01
2       2.5     1.2146021E-01
```

2	3.0	6.1510459E-02
2	5.0	5.3089437E-03
2	10.0	2.1509817E-05
2	20.0	6.3295437E-10
3	1.0	7.101262825
3	2.0	6.473853909E-01
3	5.0	8.291768415E-03
3	10.0	2.725270026E-05
3	50.0	3.72793677E-23
5	1.0	3.609605896E02
5	2.0	9.431049101
5	5.0	3.270627371E-02
5	10.0	5.754184999E-05
5	50.0	4.36718224E-23
10	1.0	1.807132899E08
10	2.0	1.624824040E05
10	5.0	9.758562829
10	10.0	1.614255300E-03
10	50.0	9.15098819E-23
20	1.0	6.294369369E22
20	2.0	5.770856853E16
20	5.0	4.827000521E08
20	10.0	1.787442782E02
20	50.0	1.70614838E-21

Modified Bessel Function In, n>=2
28 Values

2	0.2	5.0166876E-03
2	1.0	1.3574767E-01
2	2.0	6.8894844E-01
2	2.5	1.2764661
2	3.0	2.2452125
2	5.0	17.505615
2	10.0	2281.5189
2	20.0	3.9312785E07
3	1.0	2.216842492E-02
3	2.0	2.127399592E-01
3	5.0	1.033115017E01
3	10.0	1.758380717E01
3	50.0	2.67776414E20
5	1.0	2.714631560E-04
5	2.0	9.825679323E-03
5	5.0	2.157974547
5	10.0	7.771882864E02
5	50.0	2.27854831E20
10	1.0	2.752948040E-10
10	2.0	3.016963879E-07
10	5.0	4.580044419E-03
10	10.0	2.189170616E01
10	50.0	1.07159716E20
20	1.0	3.966835986E-25
20	2.0	4.310560576E-19
20	5.0	5.024239358E-11
20	10.0	1.250799736E-04
20	50.0	5.44200840E18

Legendre Polynomials
19 Values

1	0	1.0	1.224745

10	0	1.0	3.240370
20	0	1.0	4.527693
1	0	0.7071067	0.866025
10	0	0.7071067	0.373006
20	0	0.7071067	−0.874140
1	0	0.0	0.000000
10	0	0.0	−0.797435
20	0	0.0	0.797766
2	2	0.7071067	0.484123
10	2	0.7071067	−0.204789
20	2	0.7071067	0.910208
2	2	0.0	0.968246
10	2	0.0	0.804785
20	2	0.0	−0.799672
10	10	0.7071067	0.042505
20	10	0.7071067	−0.707252
10	10	0.0	1.360172
20	10	0.0	−0.853705

Jacobian Elliptic Function
20 Values

0.0	0.1	0.099833
0.0	0.2	0.19867
0.0	0.5	0.47943
0.0	1.0	0.84147
0.0	2.0	0.90930
0.5	0.1	0.099751
0.5	0.2	0.19802
0.5	0.5	0.47075
0.5	1.0	0.80300
0.5	2.0	0.99466
1.0	0.1	0.099668
1.0	0.2	0.19738
1.0	0.5	0.46212
1.0	1.0	0.76159
1.0	2.0	0.96403
1.0	4.0	0.99933
1.0	−0.2	−0.19738
1.0	−0.5	−0.46212
1.0	−1.0	−0.76159
1.0	−2.0	−0.96403

Chapter 7: Random Numbers

Chapter 7 of Numerical Recipes deals with the generation of random numbers drawn from various distributions. The first four procedures produce uniform deviates with a range of 0.0 to 1.0. ran0 is a procedure for improving the randomness of a system-supplied random number generator by shuffling the output. ran1 is a portable random number generator based on three linear congruential generators and a shuffler. ran2 contains a single congruential generator and a shuffler, and has the advantage of being somewhat faster, if less plentiful in possible output values. ran3 is another portable generator, based on a subtractive rather than a congruential method. As described in the Pascal *section of Numerical Recipes, ran1 and ran2 require 4-byte integers. Accordingly, we have adopted ran3 as the "default" random number generator for the* Pascal *programs.*

The transformation method is used to generate some non-uniform distributions. Resulting from this are routines expdev, *which gives exponentially distributed deviates, and* gasdev *for Gaussian deviates. The rejection method of producing non-uniform deviates is also discussed, and is used in* gamdev *(deviate with a gamma function distribution),* poidev *(deviate with a Poisson distribution), and* bnldev *(deviate with a binomial distribution). For generating random sequences of zeros and ones, there are two procedures,* irbit1 *and* irbit2, *both based on the 14 lowest significant bits in the seed* iseed, *but each using a different recurrence to proceed from step to step. The national Data Encryption Standard (DES) is discussed as the basis for a random number generator which we call* ran4. *DES itself is carried out by the routines* des.

$$\star \quad \star \quad \star \quad \star$$

Sample programs d7r1 to d7r4 are all really the same program, except that each calls a different random number generator (ran0 to ran3, respectively). They first draw four consecutive random numbers $X_1, \ldots, X_4$ from the generator in question. Then they treat the numbers as coordinates of a point. For example, they take (X_1, X_2) as a point in two dimensions, (X_1, X_2, X_3) as a point in three dimensions, etc. These points are inside boxes of unit dimension in their respective n-space. They may, however, be either inside or outside of the unit sphere in that space. For $n = 2, 3, 4$ we seek the probability that a point is inside the unit n-sphere. This number is easily calculated theoretically. For $n = 2$ it is $\pi/4$, for $n = 3$ it is $\pi/6$, and for $n = 4$ it is $\pi^2/32$. If the random number generator is not faulty, the points will fall within the unit n-sphere this fraction of the time, and the result should become increasingly accurate as the number of points increases. In programs d7r1 to d7r4 we have taken out a factor of 2^n for convenience, and used the random number generators as a statistical means of determining the value of π, $4\pi/3$, and $\pi^2/2$. Recall that the version of ran0 given in *Numerical Recipes* runs only under Turbo Pascal. However, you can easily modify ran0 for a different system.

```
PROGRAM d7r1 (input,output);
(* driver for routine RANO *)
(* calculates pi statistically using volume of unit n-sphere *)
(*$I MODFILE.PAS *)
CONST
   pi = 3.1415926;
TYPE
   IntegerArray3 = ARRAY [1..3] OF integer;
   RealArray3 = ARRAY [1..3] OF real;
   RealArray97 = ARRAY [1..97] OF real;
VAR
   i,j,k,idum,jpower: integer;
   x1,x2,x3,x4: real;
   iy: IntegerArray3;
   yprob: RealArray3;
   RanOY: real;
   RanOV: RealArray97;
FUNCTION fnc(x1,x2,x3,x4: real): real;
BEGIN
   fnc := sqrt(sqr(x1)+sqr(x2)+sqr(x3)+sqr(x4))
END;
FUNCTION twotoj(j: integer): integer;
BEGIN
   IF j = 0 THEN twotoj := 1
   ELSE twotoj := 2*twotoj(j-1)
END;
(*$I RANO.PAS *)
BEGIN
   idum := -1;
   FOR i := 1 TO 3 DO iy[i] := 0;
   writeln;
   writeln('volume of unit n-sphere, n = 2,3,4');
   writeln('# points','     pi    ',','  (4/3)*pi ',',' (1/2)*pi^2 ');
   writeln;
   FOR j := 1 TO 13 DO BEGIN
      FOR k := twotoj(j-1) TO twotoj(j) DO BEGIN
         x1 := ran0(idum);
         x2 := ran0(idum);
         x3 := ran0(idum);
         x4 := ran0(idum);
         IF fnc(x1,x2,0.0,0.0) < 1.0 THEN  iy[1] := iy[1]+1;
         IF fnc(x1,x2,x3,0.0) < 1.0 THEN  iy[2] := iy[2]+1;
         IF fnc(x1,x2,x3,x4) < 1.0 THEN  iy[3] := iy[3]+1
      END;
      jpower := twotoj(j);
      yprob[1] := 4.0*iy[1]/jpower;
      yprob[2] := 8.0*iy[2]/jpower;
      yprob[3] := 16.0*iy[3]/jpower;
      writeln(jpower:6,yprob[1]:12:6,yprob[2]:12:6,yprob[3]:12:6)
   END;
   writeln;
   writeln('actual',pi:12:6,(4.0*pi/3.0):12:6,(0.5*sqr(pi)):12:6)
```

```
END.

PROGRAM d7r2 (input,output);
(* driver for routine RAN1 *)
(* calculates pi statistically using volume of unit n-sphere *)
(*$I MODFILE.PAS *)
CONST
   pi = 3.1415926;
TYPE
   IntegerArray3 = ARRAY [1..3] OF integer;
   RealArray3 = ARRAY [1..3] OF real;
   RealArray97 = ARRAY [1..97] OF real;
VAR
   i,idum,j,k,jpower: integer;
   x1,x2,x3,x4: real;
   iy: IntegerArray3;
   yprob: RealArray3;
   Ran1Ix1,Ran1Ix2,Ran1Ix3: longint;
   Ran1R: RealArray97;
FUNCTION fnc(x1,x2,x3,x4: real): real;
BEGIN
   fnc := sqrt(sqr(x1)+sqr(x2)+sqr(x3)+sqr(x4))
END;
FUNCTION twotoj(j: integer): integer;
BEGIN
   IF j = 0 THEN twotoj := 1
   ELSE twotoj := 2*twotoj(j-1)
END;
(*$I RAN1.PAS *)
BEGIN
   idum := -1;
   FOR i := 1 TO 3 DO iy[i] := 0;
   writeln;
   writeln('volume of unit n-sphere, n = 2,3,4');
   writeln('# points','     pi    ','  (4/3)*pi ',' (1/2)*pi^2 ');
   writeln;
   FOR j := 1 TO 13 DO BEGIN
      FOR k := twotoj(j-1) TO twotoj(j) DO BEGIN
         x1 := ran1(idum);
         x2 := ran1(idum);
         x3 := ran1(idum);
         x4 := ran1(idum);
         IF fnc(x1,x2,0.0,0.0) < 1.0 THEN  iy[1] := iy[1]+1;
         IF fnc(x1,x2,x3,0.0) < 1.0 THEN  iy[2] := iy[2]+1;
         IF fnc(x1,x2,x3,x4) < 1.0 THEN  iy[3] := iy[3]+1
      END;
      jpower := twotoj(j);
      yprob[1] := 4.0*iy[1]/jpower;
      yprob[2] := 8.0*iy[2]/jpower;
      yprob[3] := 16.0*iy[3]/jpower;
      writeln(jpower:6,yprob[1]:12:6,yprob[2]:12:6,yprob[3]:12:6)
   END;
   writeln;
   writeln('actual',pi:12:6,(4.0*pi/3.0):12:6,(0.5*sqr(pi)):12:6)
END.
```

```
PROGRAM d7r3 (input,output);
(* driver for routine RAN2 *)
(* calculates pi statistically using volume of unit n-sphere *)
(*$I MODFILE.PAS *)
CONST
   pi = 3.1415926;
TYPE
   IntegerArray3 = ARRAY [1..3] OF integer;
   RealArray3 = ARRAY [1..3] OF real;
   IntegerArray97 = ARRAY [1..97] OF longint;
VAR
   i,j,k,jpower: integer;
   x1,x2,x3,x4: real;
   iy: IntegerArray3;
   yprob: RealArray3;
   idum,Ran2Iy: longint;
   Ran2Ir: IntegerArray97;
FUNCTION fnc(x1,x2,x3,x4: real): real;
BEGIN
   fnc := sqrt(sqr(x1)+sqr(x2)+sqr(x3)+sqr(x4))
END;
FUNCTION twotoj(j: integer): integer;
BEGIN
   IF j = 0 THEN twotoj := 1
   ELSE twotoj := 2*twotoj(j-1)
END;
(*$I RAN2.PAS *)
BEGIN
   idum := -1;
   FOR i := 1 TO 3 DO iy[i] := 0;
   writeln;
   writeln('volume of unit n-sphere, n := 2,3,4');
   writeln('# points','     pi     ',' (4/3)*pi ',' (1/2)*pi^2 ');
   writeln;
   FOR j := 1 TO 13 DO BEGIN
      FOR k := twotoj(j-1) TO twotoj(j) DO BEGIN
         x1 := ran2(idum);
         x2 := ran2(idum);
         x3 := ran2(idum);
         x4 := ran2(idum);
         IF fnc(x1,x2,0.0,0.0) < 1.0 THEN  iy[1] := iy[1]+1;
         IF fnc(x1,x2,x3,0.0) < 1.0 THEN  iy[2] := iy[2]+1;
         IF fnc(x1,x2,x3,x4) < 1.0 THEN  iy[3] := iy[3]+1
      END;
      jpower := twotoj(j);
      yprob[1] := 4.0*iy[1]/jpower;
      yprob[2] := 8.0*iy[2]/jpower;
      yprob[3] := 16.0*iy[3]/jpower;
      writeln(jpower:6,yprob[1]:12:6,yprob[2]:12:6,yprob[3]:12:6)
   END;
   writeln;
   writeln('actual',pi:12:6,(4.0*pi/3.0):12:6,(0.5*sqr(pi)):12:6)
END.

PROGRAM d7r4 (input,output);
(* driver for routine RAN3 *)
(* calculates pi statistically using volume of unit n-sphere *)
```

```
(*$I MODFILE.PAS *)
CONST
   pi = 3.1415926;
TYPE
   IntegerArray3 = ARRAY [1..3] OF integer;
   RealArray3 = ARRAY [1..3] OF real;
   RealArray55 = ARRAY [1..55] OF real;
VAR
   i,idum,j,k,jpower: integer;
   x1,x2,x3,x4: real;
   iy: IntegerArray3;
   yprob: RealArray3;
   Ran3Inext,Ran3Inextp: integer;
   Ran3Ma: RealArray55;
FUNCTION fnc(x1,x2,x3,x4: real): real;
BEGIN
   fnc := sqrt(sqr(x1)+sqr(x2)+sqr(x3)+sqr(x4))
END;
FUNCTION twotoj(j: integer): integer;
BEGIN
   IF j = 0 THEN twotoj := 1
   ELSE twotoj := 2*twotoj(j-1)
END;
(*$I RAN3.PAS *)
BEGIN
   idum := -1;
   FOR i := 1 TO 3 DO iy[i] := 0;
   writeln;
   writeln('volume of unit n-sphere, n := 2,3,4');
   writeln('# points','      pi      ',' (4/3)*pi ',' (1/2)*pi^2 ');
   writeln;
   FOR j := 1 TO 13 DO BEGIN
      FOR k := twotoj(j-1) TO twotoj(j) DO BEGIN
         x1 := ran3(idum);
         x2 := ran3(idum);
         x3 := ran3(idum);
         x4 := ran3(idum);
         IF fnc(x1,x2,0.0,0.0) < 1.0 THEN  iy[1] := iy[1]+1;
         IF fnc(x1,x2,x3,0.0) < 1.0 THEN  iy[2] := iy[2]+1;
         IF fnc(x1,x2,x3,x4) < 1.0 THEN  iy[3] := iy[3]+1
      END;
      FOR i := 1 TO 3 DO BEGIN
         jpower := twotoj(j);
         yprob[1] := 4.0*iy[1]/jpower;
         yprob[2] := 8.0*iy[2]/jpower;
         yprob[3] := 16.0*iy[3]/jpower
      END;
      writeln(jpower:6,yprob[1]:12:6,yprob[2]:12:6,yprob[3]:12:6)
   END;
   writeln;
   writeln('actual',pi:12:6,(4.0*pi/3.0):12:6,(0.5*sqr(pi)):12:6)
END.
```

Routine **expdev** generates random numbers drawn from an exponential deviate. Sample program d7r5 makes one thousand calls to **expdev** and bins the results into 21 bins, the contents of which are tallied in array x[i]. Then the sum **total** of all

bins is taken, since some of the numbers will be too large to have fallen in any of the
bins. The x[i] are scaled to `total`, and then compared to a similarly normalized
exponential which is called `expect`.

```pascal
PROGRAM d7r5 (input,output);
(* driver for routine EXPDEV *)
(*$I MODFILE.PAS *)
CONST
   npts = 1000;
   ee = 2.718281828;
TYPE
   RealArray21 = ARRAY [0..20] OF real;
   RealArray55 = ARRAY [1..55] OF real;
VAR
   i,idum,j: integer;
   expect,total,y: real;
   trig,x: RealArray21;
   Ran3Inext,Ran3Inextp: integer;
   Ran3Ma: RealArray55;
(*$I RAN3.PAS *)
(*$I EXPDEV.PAS *)
BEGIN
   FOR i := 0 TO 20 DO BEGIN
      trig[i] := i/20.0;
      x[i] := 0.0
   END;
   idum := -1;
   FOR i := 1 TO npts DO BEGIN
      y := expdev(idum);
      FOR j := 1 TO 20 DO
         IF (y < trig[j]) AND (y > trig[j-1]) THEN
            x[j] := x[j]+1.0
   END;
   total := 0.0;
   FOR i := 1 TO 20 DO total := total+x[i];
   writeln;
   writeln('exponential distribution with',npts:7,' points');
   writeln('   interval','      observed','      expected');
   writeln;
   FOR i := 1 TO 20 DO BEGIN
      x[i] := x[i]/total;
      expect := exp(-(trig[i-1]+trig[i])/2.0);
      expect := expect*0.05*ee/(ee-1);
      writeln(trig[i-1]:6:2,trig[i]:6:2,x[i]:12:6,expect:12:6)
   END
END.
```

`gasdev` generates random numbers from a Gaussian deviate. Example d7r6 takes
one thousand of these and puts them into 21 bins. For the purpose of binning, the
center of the Gaussian is shifted over by `nover2=10` bins, to put it in the middle
bin. The remainder of the program simply plots the contents of the bins, to illustrate
that they have the characteristic Gaussian bell-shape. This allows a quick, though
superficial, check of the integrity of the routine.

```pascal
PROGRAM d7r6 (input,output);
(* driver for routine GASDEV *)
(*$I MODFILE.PAS *)
```

```
CONST
   n = 20;
   nover2 = 10;     (* n/2 *)
   npts = 1000;
   iscal = 400;
   llen = 50;
TYPE
   CharArray50 = ARRAY [1..50] OF char;
   RealArray55 = ARRAY [1..55] OF real;
VAR
   i,idum,iset,j,k,klim: integer;
   gset: real;
   words: CharArray50;
   dist: ARRAY [-nover2..nover2] OF real;
   Ran3Inext,Ran3Inextp: integer;
   Ran3Ma: RealArray55;
   GasdevIset: integer;
   GasdevGset: real;
(*$I RAN3.PAS *)
(*$I GASDEV.PAS *)
BEGIN
   GasdevIset := 0;     (* initializes routine gasdev *)
   idum := -13;         (* initializes ran3 *)
   FOR j := -nover2 TO nover2 DO dist[j] := 0.0;
   FOR i := 1 TO npts DO BEGIN
      j := round(0.25*n*gasdev(idum));
      IF (j >= -nover2) AND (j <= nover2) THEN dist[j] := dist[j]+1
   END;
   writeln('normally distributed deviate of ',npts:6,' points');
   writeln('x':5,'p(x)':10,'graph:':9);
   FOR j := -nover2 TO nover2 DO BEGIN
      dist[j] := dist[j]/npts;
      FOR k := 1 TO 50 DO words[k] := ' ';
      klim := trunc(iscal*dist[j]);
      IF klim > llen THEN klim := llen;
      FOR k := 1 TO klim DO words[k] := '*';
      write(j/(0.25*n):8:4,dist[j]:8:4,' ');
      FOR k := 1 TO 50 DO write(words[k]);
      writeln
   END
END.
```

The next three sample programs d7r7 to d7r9 are identical to the previous one, but each drives a different random number generator and produces a different graph. d7r7 drives gamdev and displays a gamma distribution of order ia specified by the user. d7r8 drives poidev and produces a Poisson distribution with mean xm specified by the user. d7r9 drives bnldev and produces a binomial distribution, also with specified xm.

```
PROGRAM d7r7 (input,output);
(* driver for routine GAMDEV *)
(*$I MODFILE.PAS *)
LABEL 99;
CONST
   n = 20;
   npts = 1000;
```

```
      iscal = 200;
      llen = 50;
TYPE
      CharArray50 = ARRAY [1..50] OF char;
      RealArray21 = ARRAY [0..20] OF real;
      RealArray55 = ARRAY [1..55] OF real;
VAR
      i,ia,idum,j,k,klim: integer;
      words: CharArray50;
      dist: RealArray21;
      Ran3Inext,Ran3Inextp: integer;
      Ran3Ma: RealArray55;
(*$I RAN3.PAS *)
(*$I GAMDEV.PAS *)
BEGIN
      idum := -13;
      WHILE true DO BEGIN
         FOR j := 0 TO 20 DO dist[j] := 0.0;
         REPEAT
            writeln('Select order of Gamma distribution (n=1..20), -1 to end --');
            readln(ia);
         UNTIL ia <= 20;
         IF ia < 0 THEN GOTO 99;
         FOR i := 1 TO npts DO BEGIN
            j := trunc(gamdev(ia,idum));
            IF (j >= 0) AND (j <= 20) THEN dist[j] := dist[j]+1
         END;
         writeln;
         writeln('gamma-distribution deviate, order ',
            ia:2,' of ',npts:6,' points');
         writeln('x':6,'p(x)':7,'graph:':9);
         FOR j := 0 TO 19 DO BEGIN
            dist[j] := dist[j]/npts;
            FOR k := 1 TO 50 DO words[k] := ' ';
            klim := trunc(iscal*dist[j]);
            IF klim > llen THEN klim := llen;
            FOR k := 1 TO klim DO words[k] := '*';
            write(j:6,dist[j]:8:4,'  ');
            FOR k := 1 TO klim DO write(words[k]);
            writeln
         END
      END;
99:
END.


PROGRAM d7r8 (input,output);
(* driver for routine POIDEV *)
(*$I MODFILE.PAS *)
LABEL 99;
CONST
      n = 20;
      npts = 1000;
      iscal = 200;
      llen = 50;
TYPE
      RealArray21 = ARRAY [0..20] OF real;
      RealArray55 = ARRAY [1..55] OF real;
```

```
    CharArray50 = PACKED ARRAY [1..50] OF char;
VAR
    i,idum,j,k,klim: integer;
    xm: real;
    dist: RealArray21;
    txt: CharArray50;
    PoidevOldm,PoidevSq,PoidevAlxm,PoidevG: real;
    Ran3Inext,Ran3Inextp: integer;
    Ran3Ma: RealArray55;
(*$I RAN3.PAS *)
(*$I GAMMLN.PAS *)
(*$I POIDEV.PAS *)
BEGIN
    PoidevOldm := -1.0;    (* initializes routine poidev *)
    idum := -13;
    WHILE true DO BEGIN
        FOR j := 0 TO 20 DO dist[j] := 0.0;
        REPEAT
          writeln('Mean of Poisson distribution (x := 0.0 to 20.0); neg. to end');
          readln(xm);
        UNTIL xm <= 20.0;
        IF xm < 0.0 THEN GOTO 99;
        FOR i := 1 TO npts DO BEGIN
            j := trunc(poidev(xm,idum));
            IF (j >= 0) AND (j <= 20) THEN dist[j] := dist[j]+1
        END;
        writeln('Poisson-distributed deviate, mean ',xm:5:2,
            ' of ',npts:6,' points');
        writeln('x':5,'p(x)':8,'graph:':10);
        FOR j := 0 TO 19 DO BEGIN
            dist[j] := dist[j]/npts;
            FOR k := 1 TO 50 DO txt[k] := ' ';
            klim := trunc(iscal*dist[j]);
            IF klim > llen THEN klim := llen;
            FOR k := 1 TO klim DO txt[k] := '*';
            writeln(1.0*j:6:2,dist[j]:8:4,'   ',txt)
        END
    END;
99:
END.


PROGRAM d7r9(input,output);
(* driver for routine BNLDEV *)
(*$I MODFILE.PAS *)
LABEL 99;
CONST
    n = 20;
    npts = 1000;
    iscal = 200;
    nn = 100;
TYPE
    RealArray55 = ARRAY [1..55] OF real;
    RealArray21 = ARRAY [0..20] OF real;
    CharArray50 = PACKED ARRAY [1..50] OF char;
VAR
    Ran3Inext,Ran3Inextp: integer;
    Ran3Ma: RealArray55;
```

```
      BnldevNold: integer;
      BnldevPold: real;
      BnldevOldg: real;
      BnldevEn: real;
      BnldevPc: real;
      BnldevPlog: real;
      BnldevPclog: real;
      i,j,k,idum,klim,llen: integer;
      pp,xm: real;
      dist: RealArray21;
      txt: CharArray50;
(*$I RAN3.PAS *)
(*$I GAMMLN.PAS *)
(*$I BNLDEV.PAS *)
BEGIN
      idum := -133;
      BnldevNold := -1;
      BnldevPold := -1.0;
      llen := 50;
      WHILE true DO BEGIN
         FOR j := 0 TO 20 DO dist[j] := 0.0;
         writeln('Mean of binomial distribution (0 to 20) (neg to end)');
         read(xm);
         IF xm < 0 THEN GOTO 99;
         pp := xm/nn;
         FOR i := 1 TO npts DO BEGIN
            j := round(bnldev(pp,nn,idum));
            IF (j >= 0) AND (j <= 20) THEN dist[j] := dist[j]+1
         END;
         writeln('x':4,'p(x)':8,'graph:':10);
         FOR j := 0 TO 19 DO BEGIN
            FOR k := 1 TO llen DO txt[k] := ' ';
            dist[j] := dist[j]/npts;
            klim := round(iscal*dist[j])+1;
            IF klim > llen THEN klim := llen;
            FOR k := 1 TO klim DO txt[k] := '*';
            writeln(j:4,dist[j]:9:4,'   ',txt)
         END
      END;
99:
END.
```

Procedures `irbit1` and `irbit2` both generate random series of ones and zeros. The sample programs d7r10 and d7r11 for the two are the same, and they check that the series have correct statistical properties (or more exactly, that they have at least one correct property). They look for a 1 in the series and count how many zeros follow it before the next 1 appears. The result is stored as a distribution. There should be, for example, a 50% chance of no zeros, a 25% chance of exactly one zero, and so on.

```
PROGRAM d7r10 (input,output);
(* driver for routine IRBIT1 *)
(* calculate distribution of runs of zeros *)
(*$I MODFILE.PAS *)
CONST
   nbin = 15;
```

```
   ntries = 500;
TYPE
   RealArrayNBIN = ARRAY [1..nbin] OF real;
VAR
   i,iflg,ipts,iseed,j,n,idum: integer;
   twoinv: real;
   delay: RealArrayNBIN;
(*$I IRBIT1.PAS *)
BEGIN
   iseed := 1234;
   FOR i := 1 TO nbin DO delay[i] := 0.0;
   writeln('distribution of runs of n zeros');
   writeln('n':6,'probability':22,'expected':18);
   ipts := 0;
   FOR i := 1 TO ntries DO BEGIN
      IF irbit1(iseed) = 1 THEN BEGIN
         ipts := ipts+1;
         iflg := 0;
         FOR j := 1 TO nbin DO BEGIN
            idum := irbit1(iseed);
            IF (idum = 1) AND (iflg = 0) THEN BEGIN
               iflg := 1;
               delay[j] := delay[j]+1.0
            END
         END
      END
   END;
   twoinv := 0.5;
   FOR n := 1 TO nbin DO BEGIN
      writeln((n-1):6,(delay[n]/ipts):19:4,twoinv:20:4);
      twoinv := twoinv/2.0
   END
END.

PROGRAM d7r11(input,output);
(* driver for routine IRBIT2 *)
(* calculate distribution of runs of zeros *)
(*$I MODFILE.PAS *)
CONST
   nbin = 15;
   ntries = 500;
TYPE
   RealArrayNBIN = ARRAY [1..nbin] OF real;
VAR
   i,iflg,ipts,iseed,j,n,idum: integer;
   delay: RealArrayNBIN;
FUNCTION twoton(n: integer): integer;
BEGIN
   IF n = 0 THEN twoton := 1
   ELSE twoton := 2*twoton(n-1)
END;
(*$I IRBIT2.PAS *)
BEGIN
   iseed := 111;
   FOR i := 1 TO nbin DO delay[i] := 0.0;
   ipts := 0;
   FOR i := 1 TO ntries DO BEGIN
```

```
         IF irbit2(iseed) = 1 THEN BEGIN
            ipts := ipts+1;
            iflg := 0;
            FOR j := 1 TO nbin DO BEGIN
               idum := irbit2(iseed);
               IF (idum = 1) AND (iflg = 0) THEN BEGIN
                  iflg := 1;
                  delay[j] := delay[j]+1.0
               END
            END
         END
      END;
      writeln('distribution of runs of n zeros');
      writeln('n':6,'probability':22,'expected':18);
      FOR n := 1 TO nbin DO
         writeln((n-1):6,delay[n]/ipts:19:4,1/twoton(n):20:4);
END.
```

The next routine, **ran4**, is a random number generator with a uniform deviate, based on the data encryption standard **des**. When applied to **ran4**, the routine we used to demonstrate **ran0** to **ran3** is outrageously time consuming. **ran4** is random but very slow. To try out **ran4** we simply list the first ten random numbers for a given seed **idum=-123**. Compare your results to these; they should be exactly the same. We also generate 50 more random numbers and find their average and variance with **avevar**. Note that **ran4** runs only under Turbo Pascal.

```
PROGRAM d7r12(input,output,infile);
(* driver for routine RAN4 *)
(*$I MODFILE.PAS *)
CONST
   npt=50;
TYPE
   ByteArray32 = ARRAY [1..32] OF byte;
   ByteArray48 = ARRAY [1..48] OF byte;
   ByteArray56 = ARRAY [1..56] OF byte;
   ByteArray64 = ARRAY [1..64] OF byte;
   Immense = RECORD
      l,r: longint
   END;
   Great = RECORD
      l,c,r: word
   END;
   RealArray65 = ARRAY [1..65] OF real;
   RealArrayNP = ARRAY [1..npt] OF real;
VAR
   DesBit : ARRAY [1..32] OF longint;
   DesIp,DesIpm: ByteArray64;
   DesKns: ARRAY [1..16] OF Great;
   DesFlg: boolean;
   CyfunIet: ByteArray48;
   CyfunIpp: ByteArray32;
   CyfunIs: Array[1..16,1..4,1..8] OF integer;
   CyfunIbin: Array[0..15] OF longint;
   CyfunFlg: boolean;
   KsIcd: Immense;
   KsIpc1: ByteArray56;
```

```
   KsIpc2: ByteArray48;
   KsFlg: boolean;
   Ran4Newkey: boolean;
   Ran4Inp,Ran4Key: Immense;
   Ran4Pow: RealArray65;
   infile : text;
   ave,vrnce : real;
   idum,j : integer;
   y : RealArrayNP;
(*$I AVEVAR.PAS *)
(*$I DES.PAS *)
(*$I RAN4.PAS *)
BEGIN
   KsFlg := true;
   CyfunFlg := true;
   DesFlg := true;
   idum := -123;
   ave := 0.0;
   writeln('First 10 Random Numbers with idum = ',idum:5);
   writeln;
   writeln('#':4,'RAN4':11);
   FOR j := 1 to 10 DO y[j] := ran4(idum);
   FOR j := 1 to 10 DO writeln(j:4,y[j]:12:6);
   writeln;
   writeln('Average and Variance of Next ',npt:3);
   FOR j := 1 to npt DO y[j] := ran4(idum);
   avevar(y,npt,ave,vrnce);
   writeln;
   writeln('Average: ',ave:10:4);
   writeln('Variance:',vrnce:10:4);
   writeln;
   writeln('Expected Result for an Infinite Sample:');
   writeln;
   writeln('Average: ',0.5:10:4);
   writeln('Variance:',1.0/12.0:10:4)
END.
```

```
         First 10 random numbers with idum = -123
                  1    0.450727
                  2    0.471781
                  3    0.758399
                  4    0.667585
                  5    0.040054
                  6    0.983527
                  7    0.552085
                  8    0.104302
                  9    0.687392
                 10    0.346837
        Average and Variance of next   50
                Average:      .5171
                Variance:     .0870
```

des is a software implementation of the national data encryption standard. The complete formal test for this standard, though long and detailed, is included in program d7r13. This test consists of feeding in a long series of input codes and checking the output for agreement with expected output codes. The input-output pairs comprising the test are contained in file `destst.dat`, listed in the Appendix to this chapter. To save you the job of comparing the many 16-character strings for accuracy, we have ended each line with the phrase "o.k." or "wrong", depending on the outcome. Once again, our implementation is only in Turbo Pascal.

```pascal
PROGRAM d7r13 (input,output,dfile);
(* driver for routine DES, Turbo Pascal version *)
(*$I MODFILE.PAS *)
LABEL 99;
TYPE
   ByteArray32 = ARRAY [1..32] OF byte;
   ByteArray48 = ARRAY [1..48] OF byte;
   ByteArray56 = ARRAY [1..56] OF byte;
   ByteArray64 = ARRAY [1..64] OF byte;
   Immense = RECORD
      l,r: longint
   END;
   Great = RECORD
      l,c,r: word
   END;
VAR
   DesBit: ARRAY [1..32] OF longint;
   DesIp,DesIpm: ByteArray64;
   DesKns: ARRAY [1..16] OF Great;
   DesFlg: boolean;
   CyfunIet: ByteArray48;
   CyfunIpp: ByteArray32;
   CyfunIs: Array[1..16,1..4,1..8] OF integer;
   CyfunIbin: Array[0..15] OF longint;
   CyfunFlg: boolean;
   KsIcd: Immense;
   KsIpc1: ByteArray56;
   KsIpc2: ByteArray48;
   KsFlg: boolean;
   idirec,i,j,m,mm,nciphr: integer;
   newkey: boolean;
   iin,iout,key: Immense;
   hin,hkey,hout,hcmp: string[17];
   verdct: string[8];
   txt: string[60];
   txt2: string[6];
   dfile,infile: text;
(*$I DES.PAS *)
FUNCTION hex2longint(ch: char): longint;
(* Coverts character representing hexadecimal number to its
long integer value in a machine-independent way. *)
BEGIN
   IF (ch >= '0') AND (ch <= '9') THEN
      hex2longint := ord(ch)-ord('0')
   ELSE
      hex2longint := ord(ch)-ord('A')+10
END;
FUNCTION longint2hex(i: longint): char;
(* Inverse of hex2int *)
BEGIN
   IF i <= 9 THEN
      longint2hex := chr(i+ord('0'))
```

```
      ELSE
         longint2hex := chr(i-10+ord('A'))
END;
PROCEDURE reverse(VAR input: Immense);
(* Reverse bits of type Immense *)
VAR
      temp: Immense;
      i: integer;
BEGIN
      temp.r := 0;
      temp.l := 0;
      FOR i := 1 TO 32 DO BEGIN
         temp.r := (temp.r SHL 1) OR (input.l AND $1);
         temp.l := (temp.l SHL 1) OR (input.r AND $1);
         input.r := input.r SHR 1;
         input.l := input.l SHR 1
      END;
      input.r := temp.r;
      input.l := temp.l
END;
BEGIN
      DesFlg := true;
      CyfunFlg := true;
      KsFlg := true;
      NROpen(dfile,'destst.dat');
      readln(dfile,txt);
      writeln(txt);
      WHILE true DO BEGIN
         readln(dfile,txt);
         IF eof(dfile) THEN GOTO 99;
         writeln(txt);
         readln(dfile,nciphr);
         readln(dfile,txt2);
         IF txt2 = 'encode' THEN idirec := 0;
         IF txt2 = 'decode' THEN idirec := 1;
         REPEAT
            writeln('key':10,'plaintext':20,
               'expected cipher':21,'actual cipher':15);
            mm := 16;
            IF nciphr < 16 THEN mm := nciphr;
            nciphr := nciphr-16;
            FOR m := 1 TO mm DO BEGIN
               readln(dfile,hkey,hin,hcmp);
               iin.l := 0;
               iin.r := 0;
               key.l := 0;
               key.r := 0;
               FOR i := 2 TO 9 DO BEGIN
                  j := i+8;
                  iin.l := (iin.l SHL 4) OR hex2longint(hin[i]);
                  key.l := (key.l SHL 4) OR hex2longint(hkey[i]);
                  iin.r := (iin.r SHL 4) OR hex2longint(hin[j]);
                  key.r := (key.r SHL 4) OR hex2longint(hkey[j])
               END;
               newkey := true;
               reverse(iin);
               reverse(key);
```

```
            des(iin,key,newkey,idirec,iout);
            reverse(iout);
            hout := '                    ';
            FOR i := 9 DOWNTO 2 DO BEGIN
               j := i+8;
               hout[i] := longint2hex(iout.l AND $F);
               hout[j] := longint2hex(iout.r AND $F);
               iout.l := iout.l SHR 4;
               iout.r := iout.r SHR 4
            END;
            verdct := 'wrong   ';
            IF hcmp = hout THEN verdct := 'o.k.    ';
            writeln(hkey,hin,hcmp,hout,verdct:10)
         END;
         writeln('press RETURN to continue ...');
         readln
      UNTIL nciphr <= 0
   END;
99:
END.
```

Appendix

File destst.dat:

```
DES Validation, as per NBS publication 500-20
*** Initial Permutation and Expansion test: ***
      Key            Plaintext      Expected Cipher
  0101010101010101 95F8A5E5DD31D900 8000000000000000
  0101010101010101 DD7F121CA5015619 4000000000000000
  0101010101010101 2E8653104F3834EA 2000000000000000
  0101010101010101 4BD388FF6CD81D4F 1000000000000000
  0101010101010101 20B9E767B2FB1456 0800000000000000
  0101010101010101 55579380D77138EF 0400000000000000
  0101010101010101 6CC5DEFAAF04512F 0200000000000000
  0101010101010101 0D9F279BA5D87260 0100000000000000
  0101010101010101 D9031B0271BD5A0A 0080000000000000
  0101010101010101 424250B37C3DD951 0040000000000000
  0101010101010101 B8061B7ECD9A21E5 0020000000000000
  0101010101010101 F15D0F286B65BD28 0010000000000000
  0101010101010101 ADD0CC8D6E5DEBA1 0008000000000000
  0101010101010101 E6D5F82752AD63D1 0004000000000000
  0101010101010101 ECBFE3BD3F591A5E 0002000000000000
  0101010101010101 F356834379D165CD 0001000000000000
  0101010101010101 2B9F982F20037FA9 0000800000000000
  0101010101010101 889DE068A16F0BE6 0000400000000000
  0101010101010101 E19E275D846A1298 0000200000000000
  0101010101010101 329A8ED523D71AEC 0000100000000000
  0101010101010101 E7FCE22557D23C97 0000080000000000
  0101010101010101 12A9F5817FF2D65D 0000040000000000
  0101010101010101 A484C3AD38DC9C19 0000020000000000
  0101010101010101 FBE00A8A1EF8AD72 0000010000000000
  0101010101010101 750D079407521363 0000008000000000
  0101010101010101 64FEED9C724C2FAF 0000004000000000
  0101010101010101 F02B263B328E2B60 0000002000000000
  0101010101010101 9D64555A9A10B852 0000001000000000
  0101010101010101 D106FF0BED5255D7 0000000800000000
```

```
0101010101010101 E1652C6B138C64A5 0000000400000000
0101010101010101 E428581186EC8F46 0000000200000000
0101010101010101 AEB5F5EDE22D1A36 0000000100000000
0101010101010101 E943D7568AEC0C5C 0000000080000000
0101010101010101 DF98C8276F54B04B 0000000040000000
0101010101010101 B160E4680F6C696F 0000000020000000
0101010101010101 FA0752B07D9C4AB8 0000000010000000
0101010101010101 CA3A2B036DBC8502 0000000008000000
0101010101010101 5E0905517BB59BCF 0000000004000000
0101010101010101 814EEB3B91D90726 0000000002000000
0101010101010101 4D49DB1532919C9F 0000000001000000
0101010101010101 25EB5FC3F8CF0621 0000000000800000
0101010101010101 AB6A20C0620D1C6F 0000000000400000
0101010101010101 79E90DBC98F92CCA 0000000000200000
0101010101010101 866ECEDD8072BB0E 0000000000100000
0101010101010101 8B54536F2F3E64A8 0000000000080000
0101010101010101 EA51D3975595B86B 0000000000040000
0101010101010101 CAFFC6AC4542DE31 0000000000020000
0101010101010101 8DD45A2DDF90796C 0000000000010000
0101010101010101 1029D55E880EC2D0 0000000000008000
0101010101010101 5D86CB23639DBEA9 0000000000004000
0101010101010101 1D1CA853AE7C0C5F 0000000000002000
0101010101010101 CE332329248F3228 0000000000001000
0101010101010101 8405D1ABE24FB942 0000000000000800
0101010101010101 E643D78090CA4207 0000000000000400
0101010101010101 48221B9937748A23 0000000000000200
0101010101010101 DD7C0BBD61FAFD54 0000000000000100
0101010101010101 2FBC291A570DB5C4 0000000000000080
0101010101010101 E07C30D7E4E26E12 0000000000000040
0101010101010101 0953E2258E8E90A1 0000000000000020
0101010101010101 5B711BC4CEEBF2EE 0000000000000010
0101010101010101 CC083F1E6D9E85F6 0000000000000008
0101010101010101 D2FD8867D50D2DFE 0000000000000004
0101010101010101 06E7EA22CE92708F 0000000000000002
0101010101010101 166B40B44ABA4BD6 0000000000000001
```

*** Inverse Permutation and Expansion test ***

```
      Key            Plaintext        Expected Cipher
0101010101010101 8000000000000000 95F8A5E5DD31D900
0101010101010101 4000000000000000 DD7F121CA5015619
0101010101010101 2000000000000000 2E8653104F3834EA
0101010101010101 1000000000000000 4BD388FF6CD81D4F
0101010101010101 0800000000000000 20B9E767B2FB1456
0101010101010101 0400000000000000 55579380D77138EF
0101010101010101 0200000000000000 6CC5DEFAAF04512F
0101010101010101 0100000000000000 0D9F279BA5D87260
0101010101010101 0080000000000000 D9031B0271BD5A0A
0101010101010101 0040000000000000 424250B37C3DD951
0101010101010101 0020000000000000 B8061B7ECD9A21E5
0101010101010101 0010000000000000 F15D0F286B65BD28
0101010101010101 0008000000000000 ADD0CC8D6E5DEBA1
0101010101010101 0004000000000000 E6D5F82752AD63D1
0101010101010101 0002000000000000 ECBFE3BD3F591A5E
0101010101010101 0001000000000000 F356834379D165CD
0101010101010101 0000800000000000 2B9F982F20037FA9
0101010101010101 0000400000000000 889DE068A16F0BE6
0101010101010101 0000200000000000 E19E275D846A1298
0101010101010101 0000100000000000 329A8ED523D71AEC
```

```
0101010101010101  0000080000000000  E7FCE22557D23C97
0101010101010101  0000040000000000  12A9F5817FF2D65D
0101010101010101  0000020000000000  A484C3AD38DC9C19
0101010101010101  0000010000000000  FBE00A8A1EF8AD72
0101010101010101  0000008000000000  750D079407521363
0101010101010101  0000004000000000  64FEED9C724C2FAF
0101010101010101  0000002000000000  F02B263B328E2B60
0101010101010101  0000001000000000  9D64555A9A10B852
0101010101010101  0000000800000000  D106FF0BED5255D7
0101010101010101  0000000400000000  E1652C6B138C64A5
0101010101010101  0000000200000000  E428581186EC8F46
0101010101010101  0000000100000000  AEB5F5EDE22D1A36
0101010101010101  0000000080000000  E943D7568AEC0C5C
0101010101010101  0000000040000000  DF98C8276F54B04B
0101010101010101  0000000020000000  B160E4680F6C696F
0101010101010101  0000000010000000  FA0752B07D9C4AB8
0101010101010101  0000000008000000  CA3A2B036DBC8502
0101010101010101  0000000004000000  5E0905517BB59BCF
0101010101010101  0000000002000000  814EEB3B91D90726
0101010101010101  0000000001000000  4D49DB1532919C9F
0101010101010101  0000000000800000  25EB5FC3F8CF0621
0101010101010101  0000000000400000  AB6A20C0620D1C6F
0101010101010101  0000000000200000  79E90DBC98F92CCA
0101010101010101  0000000000100000  866ECEDD8072BB0E
0101010101010101  0000000000080000  8B54536F2F3E64A8
0101010101010101  0000000000040000  EA51D3975595B86B
0101010101010101  0000000000020000  CAFFC6AC4542DE31
0101010101010101  0000000000010000  8DD45A2DDF90796C
0101010101010101  0000000000008000  1029D55E880EC2D0
0101010101010101  0000000000004000  5D86CB23639DBEA9
0101010101010101  0000000000002000  1D1CA853AE7C0C5F
0101010101010101  0000000000001000  CE332329248F3228
0101010101010101  0000000000000800  8405D1ABE24FB942
0101010101010101  0000000000000400  E643D78090CA4207
0101010101010101  0000000000000200  48221B9937748A23
0101010101010101  0000000000000100  DD7C0BBD61FAFD54
0101010101010101  0000000000000080  2FBC291A570DB5C4
0101010101010101  0000000000000040  E07C30D7E4E26E12
0101010101010101  0000000000000020  0953E2258E8E90A1
0101010101010101  0000000000000010  5B711BC4CEEBF2EE
0101010101010101  0000000000000008  CC083F1E6D9E85F6
0101010101010101  0000000000000004  D2FD8867D50D2DFE
0101010101010101  0000000000000002  06E7EA22CE92708F
0101010101010101  0000000000000001  166B40B44ABA4BD6
```

*** Key Permutation tests: ***

Key	Plaintext	Expected Cipher
8001010101010101	0000000000000000	95A8D72813DAA94D
4001010101010101	0000000000000000	0EEC1487DD8C26D5
2001010101010101	0000000000000000	7AD16FFB79C45926
1001010101010101	0000000000000000	D3746294CA6A6CF3
0801010101010101	0000000000000000	809F5F873C1FD761
0401010101010101	0000000000000000	C02FAFFEC989D1FC
0201010101010101	0000000000000000	4615AA1D33E72F10
0180010101010101	0000000000000000	2055123350C00858
0140010101010101	0000000000000000	DF3B99D6577397C8
0120010101010101	0000000000000000	31FE17369B5288C9
0110010101010101	0000000000000000	DFDD3CC64DAE1642

```
0108010101010101 0000000000000000 178C83CE2B399D94
0104010101010101 0000000000000000 50F636324A9B7F80
0102010101010101 0000000000000000 A8468EE3BC18F06D
0101800101010101 0000000000000000 A2DC9E92FD3CDE92
0101400101010101 0000000000000000 CAC09F797D031287
0101200101010101 0000000000000000 90BA680B22AEB525
0101100101010101 0000000000000000 CE7A24F350E280B6
0101080101010101 0000000000000000 882BFF0AA01A0B87
0101040101010101 0000000000000000 25610288924511C2
0101020101010101 0000000000000000 C71516C29C75D170
0101018001010101 0000000000000000 5199C29A52C9F059
0101014001010101 0000000000000000 C22F0A294A71F29F
0101012001010101 0000000000000000 EE371483714C02EA
0101011001010101 0000000000000000 A81FBD448F9E522F
0101010801010101 0000000000000000 4F644C92E192DFED
0101010401010101 0000000000000000 1AFA9A66A6DF92AE
0101010201010101 0000000000000000 B3C1CC715CB879D8
0101010180010101 0000000000000000 19D032E64AB0BD8B
0101010140010101 0000000000000000 3CFAA7A7DC8720DC
0101010120010101 0000000000000000 B7265F7F447AC6F3
0101010110010101 0000000000000000 9DB73B3C0D163F54
0101010108010101 0000000000000000 8181B65BABF4A975
0101010104010101 0000000000000000 93C9B64042EAA240
0101010102010101 0000000000000000 5570530829705592
0101010101800101 0000000000000000 8638809E878787A0
0101010101400101 0000000000000000 41B9A79AF79AC208
0101010101200101 0000000000000000 7A9BE42F2009A892
0101010101100101 0000000000000000 29038D56BA6D2745
0101010101080101 0000000000000000 5495C6ABF1E5DF51
0101010101040101 0000000000000000 AE13DBD561488933
0101010101020101 0000000000000000 024D1FFA8904E389
0101010101018001 0000000000000000 D1399712F99BF02E
0101010101014001 0000000000000000 14C1D7C1CFFEC79E
0101010101012001 0000000000000000 1DE5279DAE3BED6F
0101010101011001 0000000000000000 E941A33F85501303
0101010101010801 0000000000000000 DA99DBBC9A03F379
0101010101010401 0000000000000000 B7FC92F91D8E92E9
0101010101010201 0000000000000000 AE8E5CAA3CA04E85
0101010101010180 0000000000000000 9CC62DF43B6EED74
0101010101010140 0000000000000000 D863DBB5C59A91A0
0101010101010120 0000000000000000 A1AB2190545B91D7
0101010101010110 0000000000000000 0875041E64C570F7
0101010101010108 0000000000000000 5A594528BEBEF1CC
0101010101010104 0000000000000000 FCDB3291DE21F0C0
0101010101010102 0000000000000000 869EFD7F9F265A09
```

*** Test of right-shifts in Decryption ***

```
     Key           Plaintext        Expected Cipher
8001010101010101 95A8D72813DAA94D 0000000000000000
4001010101010101 0EEC1487DD8C26D5 0000000000000000
2001010101010101 7AD16FFB79C45926 0000000000000000
1001010101010101 D3746294CA6A6CF3 0000000000000000
0801010101010101 809F5F873C1FD761 0000000000000000
0401010101010101 C02FAFFEC989D1FC 0000000000000000
0201010101010101 4615AA1D33E72F10 0000000000000000
0180010101010101 2055123350C00858 0000000000000000
0140010101010101 DF3B99D6577397C8 0000000000000000
0120010101010101 31FE17369B5288C9 0000000000000000
```

```
0110010101010101 DFDD3CC64DAE1642 0000000000000000
0108010101010101 178C83CE2B399D94 0000000000000000
0104010101010101 50F636324A9B7F80 0000000000000000
0102010101010101 A8468EE3BC18F06D 0000000000000000
0101800101010101 A2DC9E92FD3CDE92 0000000000000000
0101400101010101 CAC09F797D031287 0000000000000000
0101200101010101 90BA680B22AEB525 0000000000000000
0101100101010101 CE7A24F350E280B6 0000000000000000
0101080101010101 882BFF0AA01A0B87 0000000000000000
0101040101010101 25610288924511C2 0000000000000000
0101020101010101 C71516C29C75D170 0000000000000000
0101018001010101 5199C29A52C9F059 0000000000000000
0101014001010101 C22F0A294A71F29F 0000000000000000
0101012001010101 EE371483714C02EA 0000000000000000
0101011001010101 A81FBD448F9E522F 0000000000000000
0101010801010101 4F644C92E192DFED 0000000000000000
0101010401010101 1AFA9A66A6DF92AE 0000000000000000
0101010201010101 B3C1CC715CB879D8 0000000000000000
0101010180010101 19D032E64AB0BD8B 0000000000000000
0101010140010101 3CFAA7A7DC8720DC 0000000000000000
0101010120010101 B7265F7F447AC6F3 0000000000000000
0101010110010101 9DB73B3C0D163F54 0000000000000000
0101010108010101 8181B65BABF4A975 0000000000000000
0101010104010101 93C9B64042EAA240 0000000000000000
0101010102010101 5570530829705592 0000000000000000
0101010101800101 8638809E878787A0 0000000000000000
0101010101400101 41B9A79AF79AC208 0000000000000000
0101010101200101 7A9BE42F2009A892 0000000000000000
0101010101100101 29038D56BA6D2745 0000000000000000
0101010101080101 5495C6ABF1E5DF51 0000000000000000
0101010101040101 AE13DBD561488933 0000000000000000
0101010101020101 024D1FFA8904E389 0000000000000000
0101010101018001 D1399712F99BF02E 0000000000000000
0101010101014001 14C1D7C1CFFEC79E 0000000000000000
0101010101012001 1DE5279DAE3BED6F 0000000000000000
0101010101011001 E941A33F85501303 0000000000000000
0101010101010801 DA99DBBC9A03F379 0000000000000000
0101010101010401 B7FC92F91D8E92E9 0000000000000000
0101010101010201 AE8E5CAA3CA04E85 0000000000000000
0101010101010180 9CC62DF43B6EED74 0000000000000000
0101010101010140 D863DBB5C59A91A0 0000000000000000
0101010101010120 A1AB2190545B91D7 0000000000000000
0101010101010110 0875041E64C570F7 0000000000000000
0101010101010108 5A594528BEBEF1CC 0000000000000000
0101010101010104 FCDB3291DE21F0C0 0000000000000000
0101010101010102 869EFD7F9F265A09 0000000000000000
*** Data permutation test: ***
      Key           Plaintext         Expected Cipher
1046913489980131 0000000000000000 88D55E54F54C97B4
1007103489988020 0000000000000000 0C0CC00C83EA48FD
10071034C8980120 0000000000000000 83BC8EF3A6570183
1046103489988020 0000000000000000 DF725DCAD94EA2E9
1086911519190101 0000000000000000 E652B53B550BE8B0
1086911519580101 0000000000000000 AF527120C485CBB0
5107B01519580101 0000000000000000 0F04CE393DB926D5
1007B01519190101 0000000000000000 C9F00FFC74079067
3107915498080101 0000000000000000 7CFD82A593252B4E
```

```
3107919498080101  0000000000000000  CB49A2F9E91363E3
10079115B9080140  0000000000000000  00B588BE70D23F56
3107911598080140  0000000000000000  406A9A6AB43399AE
1007D01589980101  0000000000000000  6CB773611DCA9ADA
910791158998O101  0000000000000000  67FD21C17DBB5D70
9107D01589190101  0000000000000000  9592CB4110430787
1007D01598980120  0000000000000000  A6B7FF68A318DDD3
1007940498190101  0000000000000000  4D102196C914CA16
0107910491190401  0000000000000000  2DFA9F4573594965
0107910491190101  0000000000000000  B46604816C0E0774
0107940491190401  0000000000000000  6E7E6221A4F34E87
19079210981A0101  0000000000000000  AA85E74643233199
1007911998190801  0000000000000000  2E5A19DB4D1962D6
10079119981A0801  0000000000000000  23A866A809D30894
1007921098190101  0000000000000000  D812D961F017D320
100791159819010B  0000000000000000  055605816E58608F
1004801598190101  0000000000000000  ABD88E8B1B7716F1
1004801598190102  0000000000000000  537AC95BE69DA1E1
1004801598190108  0000000000000000  AED0F6AE3C25CDD8
1002911498100104  0000000000000000  B3E35A5EE53E7B8D
1002911598190104  0000000000000000  61C79C71921A2EF8
1002911598100201  0000000000000000  E2F5728F0995013C
1002911698100101  0000000000000000  1AEAC39A61F0A464
```
*** S-Box test: ***
```
      Key          Plaintext      Expected Cipher
7CA110454A1A6E57  01A1D6D039776742  690F5B0D9A26939B
0131D9619DC1376E  5CD54CA83DEF57DA  7A389D10354BD271
07A1133E4A0B2686  0248D43806F67172  868EBB51CAB4599A
3849674C2602319E  51454B582DDF440A  7178876E01F19B2A
04B915BA43FEB5B6  42FD443059577FA2  AF37FB421F8C4095
0113B970FD34F2CE  059B5E0851CF143A  86A560F10EC6D85B
0170F175468FB5E6  0756D8E0774761D2  0CD3DA020021DC09
43297FAD38E373FE  762514B829BF486A  EA676B2CB7DB2B7A
07A7137045DA2A16  3BDD119049372802  DFD64A815CAF1A0F
04689104C2FD3B2F  26955F6835AF609A  5C513C9C4886C088
37D06BB516CB7546  164D5E404F275232  0A2AEEAE3FF4AB77
1F08260D1AC2465E  6B056E18759F5CCA  EF1BF03E5DFA575A
584023641ABA6176  004BD6EF09176062  88BF0DB6D70DEE56
025816164629B007  480D39006EE762F2  A1F9915541020B56
49793EBC79B3258F  437540C8698F3CFA  6FBF1CAFCFFD0556
4FB05E1515AB73A7  072D43A077075292  2F22E49BAB7CA1AC
49E95D6D4CA229BF  02FE55778117F12A  5A6B612CC26CCE4A
018310DC409B26D6  1D9D5C5018F728C2  5F4C038ED12B2E41
1C587F1C13924FEF  305532286D6F295A  63FAC0D034D9F793
```
*** End of Test ***

Chapter 8: Sorting

Chapter 8 of Numerical Recipes covers a variety of sorting tasks including sorting arrays into numerical order, preparing an index table for the order of an array, and preparing a rank table showing the rank order of each element in the array. piksrt *sorts a single array by the straight insertion method.* piksr2 *sorts by the same method but makes the corresponding rearrangement of a second array as well.* shell *carries out a Shell sort.* sort *and* sort2 *both do a Heapsort, and they are related in the same way as* piksrt *and* piksr2. *That is,* sort *sorts a single array;* sort2 *sorts an array while correspondingly rearranging a second array.* qcksrt *sorts an array by the Quicksort algorithm, which is fast (on average) but requires a small amount of auxiliary storage.*

indexx *indexes an array. That is, it produces a second array that contains pointers to the elements of the original array in the order of their size.* sort3 *uses* indexx *and illustrates its value by sorting one array while making corresponding rearrangements in two others.* rank *produces the rank table for an array of data. The rank table is a second array whose elements list the rank order of the corresponding elements of the original array.*

Finally, the routines eclass *and* eclazz *deal with equivalence classes.* eclass *gives the equivalence class of each element in an array based on a list of equivalent pairs which it is given as input.* eclazz *gives the same output but bases it on a procedure named* equiv(j,k) *which tells whether two array elements* j *and* k *are in the same equivalence class.*

⋆ ⋆ ⋆ ⋆

Routine **piksrt** sorts an array by straight insertion. Sample program d8r1 provides it with a 100-element array from file **tarray.dat** which is listed in the Appendix to this chapter. The program prints both the original and the sorted array for comparison.

```
PROGRAM d8r1(input,output,dfile);
(* driver for routine PIKSRT *)
(*$I MODFILE.PAS *)
CONST
   np = 100;
TYPE
   RealArrayNP = ARRAY [1..np] OF real;
VAR
   i,j: integer;
   a: RealArrayNP;
   dfile: text;
(*$I PIKSRT.PAS *)
```

99

```
BEGIN
    NROpen(dfile,'tarray.dat');
    readln(dfile);
    FOR i := 1 TO 100 DO read(dfile,a[i]);
    close(dfile);
(* write original array *)
    writeln('original array:');
    FOR i := 1 TO 10 DO BEGIN
        FOR j := 1 TO 10 DO write(a[10*(i-1)+j]:6:2);
        writeln
    END;
(* write sorted array *)
    piksrt(np,a);
    writeln('sorted array:');
    FOR i := 1 TO 10 DO BEGIN
        FOR j := 1 TO 10 DO write(a[10*(i-1)+j]:6:2);
        writeln
    END
END.
```

piksr2 sorts an array, and simultaneously rearranges a second array (of the same size) correspondingly. In program d8r2, the first array a[i] is again taken from tarray.dat. The second is defined by b[i]=i-1. In other words, b is originally sorted and a is not. After a call to piksr2, the situation should be reversed. With a second call, this time with b as the first argument and a as the second, the two arrays should be returned to their original form.

```
PROGRAM d8r2(input,output,dfile);
(* driver for routine PIKSR2 *)
(*$I MODFILE.PAS *)
CONST
    np = 100;
TYPE
    RealArrayNP = ARRAY [1..np] OF real;
VAR
    i,j: integer;
    a,b: RealArrayNP;
    dfile: text;
(*$I PIKSR2.PAS *)
BEGIN
    NROpen(dfile,'tarray.dat');
    readln(dfile);
    FOR i := 1 TO np DO read(dfile,a[i]);
    close(dfile);
(* generate b-array *)
    FOR i := 1 TO np DO b[i] := i-1;
(* sort a and mix b *)
    piksr2(np,a,b);
    writeln('after sorting a and mixing b, array a is:');
    FOR i := 1 TO 10 DO BEGIN
        FOR j := 1 TO 10 DO write(a[10*(i-1)+j]:6:2);
        writeln
    END;
    writeln('... and array b is:');
    FOR i := 1 TO 10 DO BEGIN
        FOR j := 1 TO 10 DO write(b[10*(i-1)+j]:6:2);
```

```
         writeln
   END;
   writeln('press return to continue ...');
   readln;
(* sort b and mix a *)
   piksr2(np,b,a);
   writeln('after sorting b and mixing a, array a is:');
   FOR i := 1 TO 10 DO BEGIN
       FOR j := 1 TO 10 DO write(a[10*(i-1)+j]:6:2);
       writeln
   END;
   writeln('... and array b is:');
   FOR i := 1 TO 10 DO BEGIN
       FOR j := 1 TO 10 DO write(b[10*(i-1)+j]:6:2);
       writeln
   END
END.
```

Procedure shell does a Shell sort of a data array. The calling format is identical to that of piksrt, and so we use the same sample program, now called d8r9.

```
PROGRAM d8r9(input,output,dfile);
(* driver for routine SHELL *)
(*$I MODFILE.PAS *)
CONST
   npt = 100;
TYPE
   RealArrayNP = ARRAY [1..npt] OF real;
VAR
   i,j: integer;
   a: RealArrayNP;
   dfile: text;
(*$I SHELL.PAS *)
BEGIN
   NROpen(dfile,'tarray.dat');
   readln(dfile);
   FOR i := 1 TO npt DO read(dfile,a[i]);
   close(dfile);
(* write original array *)
   writeln('Original array:');
   FOR i := 1 TO npt DIV 10 DO BEGIN
       FOR j := 1 TO 10 DO write(a[10*(i-1)+j]:6:2);
       writeln
   END;
(* write sorted array *)
   shell(npt,a);
   writeln;
   writeln('Sorted array:');
   FOR i := 1 TO npt DIV 10 DO BEGIN
       FOR j := 1 TO 10 DO write(a[10*(i-1)+j]:6:2);
       writeln
   END;
END.
```

By the same token, routines sort and sort2 employ the same programs as routines piksrt and piksr2, respectively. (Here they are called d8r3 and d8r4.) Both routines use the Heapsort algorithm. sort, however, works on a single array. sort2 sorts one array while making corresponding rearrangements to a second.

```pascal
PROGRAM d8r3(input,output,dfile);
(* driver for routine SORT *)
(*$I MODFILE.PAS *)
CONST
   np = 100;
TYPE
   RealArrayNP = ARRAY [1..np] OF real;
VAR
   i,j: integer;
   a: RealArrayNP;
   dfile: text;
(*$I SORT.PAS *)
BEGIN
   NROpen(dfile,'tarray.dat');
   readln(dfile);
   FOR i := 1 TO 100 DO read(dfile,a[i]);;
   close(dfile);
(* write original array *)
   writeln('original array:');
   FOR i := 1 TO 10 DO BEGIN
      FOR j := 1 TO 10 DO write(a[10*(i-1)+j]:6:2);
      writeln
   END;
(* write sorted array *)
   sort(np,a);
   writeln('sorted array:');
   FOR i := 1 TO 10 DO BEGIN
      FOR j := 1 TO 10 DO write(a[10*(i-1)+j]:6:2);
      writeln
   END
END.


PROGRAM d8r4(input,output,dfile);
(* driver for routine SORT2 *)
(*$I MODFILE.PAS *)
CONST
   np = 100;
TYPE
   RealArrayNP = ARRAY [1..np] OF real;
VAR
   i,j: integer;
   a,b: RealArrayNP;
   dfile: text;
(*$I SORT2.PAS *)
BEGIN
   NROpen(dfile,'tarray.dat');
   readln(dfile);
   FOR i := 1 TO 100 DO read(dfile,a[i]);
   close(dfile);
(* generate b-array *)
   FOR i := 1 TO 100 DO b[i] := i-1;
(* sort a and mix b *)
   sort2(100,a,b);
   writeln('after sorting a and mixing b, array a is:');
   FOR i := 1 TO 10 DO BEGIN
```

```
          FOR j := 1 TO 10 DO write(a[10*(i-1)+j]:6:2);
          writeln
      END;
      writeln('... and array b is:');
      FOR i := 1 TO 10 DO BEGIN
          FOR j := 1 TO 10 DO write(b[10*(i-1)+j]:6:2);
          writeln
      END;
      writeln('press return to continue...');
      readln;
(* sort b and mix a *)
      sort2(100,b,a);
      writeln('after sorting b and mixing a, array a is:');
      FOR i := 1 TO 10 DO BEGIN
          FOR j := 1 TO 10 DO write(a[10*(i-1)+j]:6:2);
          writeln
      END;
      writeln('... and array b is:');
      FOR i := 1 TO 10 DO BEGIN
          FOR j := 1 TO 10 DO write(b[10*(i-1)+j]:6:2);
          writeln
      END
END.
```

The procedure **indexx** generates the index array for a given input array. The index array indx[j] gives, for each j, the index of the element of the input array which will assume position j if the array is sorted. That is, for an input array a, the sorted version of a will be a[indx[j]]. To demonstrate this, sample program d8r5 produces an index for the array in tarray.dat. It then prints the array in the order a[indx[j]], j=1,..,100 for inspection.

```
PROGRAM d8r5(input,output,dfile);
(* driver for routine INDEXX *)
(*$I MODFILE.PAS *)
CONST
    np = 100;
TYPE
    RealArrayNP = ARRAY [1..np] OF real;
    IntegerArrayNP = ARRAY [1..np] OF integer;
VAR
    i,j: integer;
    a: RealArrayNP;
    indx: IntegerArrayNP;
    dfile: text;
(*$I INDEXX.PAS *)
BEGIN
    NROpen(dfile,'tarray.dat');
    readln(dfile);
    FOR i := 1 TO 100 DO read(dfile,a[i]);
    close(dfile);
(* generate index for sorted array *)
    indexx(np,a,indx);
(* write original array *)
    writeln('original array:');
    FOR i := 1 TO 10 DO BEGIN
        FOR j := 1 TO 10 DO write(a[10*(i-1)+j]:6:2);
```

```
         writeln
      END;
(* write sorted array *)
      writeln('sorted array:');
      FOR i := 1 TO 10 DO BEGIN
         FOR j := 1 TO 10 DO
            write(a[indx[10*(i-1)+j]]:6:2);
         writeln
      END
END.
```

One use for `indexx` is the management of more than two arrays. `sort3`, for example, sorts one array while making corresponding reorderings of two other arrays. In sample program `d8r6`, the first array is taken as the first 64 elements of `tarray.dat` (see Appendix). The second and third arrays are taken to be the numbers 1 to 64 in forward order and reverse order, respectively. When the first array is ordered, the second and third are scrambled, but scrambled in exactly the same way. To prove this, a text message is assigned to a character array. Then the letters are scrambled according to the order of numbers found in the rearranged second array. They are subsequently unscrambled according to the order of numbers found in the rearranged third array. If `sort3` works properly, this ought to leave the message reading in the reverse order.

```
PROGRAM d8r6(input,output,dfile);
(* driver for routine SORT3 *)
(*$I MODFILE.PAS *)
CONST
   nlen = 64;
TYPE
   CharArray40 = PACKED ARRAY [1..40] OF char;
   CharArray24 = PACKED ARRAY [1..24] OF char;
   CharArrayNLEN = PACKED ARRAY [1..nlen] OF char;
   RealArrayNP = ARRAY [1..nlen] OF real;
   IntegerArrayNP = ARRAY [1..nlen] OF integer;
VAR
   i,j: integer;
   a,b,c: RealArrayNP;
   amsg1: CharArray40;
   amsg2: CharArray24;
   n1,n2: integer;
   amsg,bmsg,cmsg: CharArrayNLEN;
   dfile: text;
(*$I INDEXX.PAS *)
(*$I SORT3.PAS *)
BEGIN
   amsg1 := 'i''d rather have a bottle in front of me ';
   n1 := 40;
   amsg2 := 'than a frontal lobotomy.';
   n2 := 24;
   FOR i := 1 TO n1 DO amsg[i] := amsg1[i];
   FOR i := 1 TO n2 DO amsg[n1+i] := amsg2[i];
   writeln;
   writeln('original message:');
   writeln(amsg);
(* read array of random numbers *)
   NRopen(dfile,'tarray.dat');
```

```
      readln(dfile);
      FOR i := 1 TO nlen DO read(dfile,a[i]);
      close(dfile);
(* create array b and array c *)
      FOR i := 1 TO nlen DO BEGIN
          b[i] := i;
          c[i] := nlen+1-i
      END;
(* sort array a while mixing ib and ic *)
      sort3(nlen,a,b,c);
(* scramble message according to array b *)
      FOR i := 1 TO nlen DO BEGIN
          j := round(b[i]);
          bmsg[i] := amsg[j]
      END;
      writeln;
      writeln('scrambled message:');
      writeln(bmsg);
(* unscramble according to array c *)
      FOR i := 1 TO nlen DO BEGIN
          j := round(c[i]);
          cmsg[j] := bmsg[i]
      END;
      writeln;
      writeln('mirrored message:');
      writeln(cmsg)
END.
```

rank is a procedure that is similar to indexx. Instead of producing an indexing array, though, it produces a rank table. For an array a[j] and rank table irank[j], entry j in irank will tell what index a[j] will have if a is sorted. irank actually takes its input information not from the array itself, but from the index array produced by indexx. Sample program d8r7 begins with the array from tarray, and feeds it to indexx and rank. The table of ranks produced is listed. To check it, the array a is copied into an array b in the rank order suggested by irank. b should then be in proper order.

```
PROGRAM d8r7(input,output,dfile);
(* driver for routine RANK *)
(*$I MODFILE.PAS *)
CONST
    np = 100;
TYPE
    IntegerArrayNP = ARRAY [1..np] OF integer;
    RealArrayNP = ARRAY [1..np] OF real;
    RealArray10 = ARRAY [1..10] OF real;
VAR
    i,j,k,l: integer;
    a: RealArrayNP;
    b: RealArray10;
    indx,irank: IntegerArrayNP;
    dfile: text;
(*$I INDEXX.PAS *)
(*$I RANK.PAS *)
BEGIN
    NROpen(dfile,'tarray.dat');
```

```
   readln(dfile);
   FOR i := 1 TO 100 DO read(dfile,a[i]);
   close(dfile);
   indexx(np,a,indx);
   rank(np,indx,irank);
   writeln('original array is:');
   FOR i := 1 TO 10 DO BEGIN
      FOR j := 1 TO 10 DO write(a[10*(i-1)+j]:6:2);
      writeln
   END;
   writeln('table of ranks is:');
   FOR i := 1 TO 10 DO BEGIN
      FOR j := 1 TO 10 DO write(irank[10*(i-1)+j]:6);
      writeln
   END;
   writeln('press return to continue...');
   readln;
   writeln('array sorted according to rank table:');
   FOR i := 1 TO 10 DO BEGIN
      FOR j := 1 TO 10 DO BEGIN
         k := 10*(i-1)+j;
         FOR l := 1 TO 100 DO
            IF irank[l] = k THEN b[j] := a[l]
      END;
      FOR j := 1 TO 10 DO write(b[j]:6:2);
   writeln
   END
END.
```

qcksrt sorts an array by the Quicksort algorithm. Its calling sequence is exactly like that of piksrt and sort, so we again rely on the same sample program, now called d8r8.

```
PROGRAM d8r8(input,output,dfile);
(* driver for routine QCKSRT *)
(*$I MODFILE.PAS *)
CONST
   np = 100;
TYPE
   RealArrayNP = ARRAY [1..np] OF real;
VAR
   i,j: integer;
   a: RealArrayNP;
   dfile: text;
(*$I QCKSRT.PAS *)
BEGIN
   NROpen(dfile,'tarray.dat');
   readln(dfile);
   FOR i := 1 TO 100 DO read(dfile,a[i]);
   close(dfile);
(* write original array *)
   writeln('original array:');
   FOR i := 1 TO 10 DO BEGIN
      FOR j := 1 TO 10 DO write(a[10*(i-1)+j]:6:2);
      writeln
   END;
(* write sorted array *)
```

```
      qcksrt(np,a);
      writeln;
      writeln('sorted array:');
      FOR i := 1 TO 10 DO BEGIN
         FOR j := 1 TO 10 DO write(a[10*(i-1)+j]:6:2);
         writeln
      END
END.
```

Procedure `eclass` generates a list of equivalence classes for the elements of an input array, based on the arrays `lista[j]` and `listb[j]` which list equivalent pairs for each j. In sample program d8r10, these lists are

$$\text{lista}: \quad 1,1,5,2,6,2,7,11,3,4,12$$
$$\text{listb}: \quad 5,9,13,6,10,14,3,7,15,8,4$$

According to these lists, 1 is equivalent to 5, 1 is equivalent to 9, etc. If you work it out, you will find the following classes:

$$\text{class1}: \quad 1,5,9,13$$
$$\text{class2}: \quad 2,6,10,14$$
$$\text{class3}: \quad 3,7,11,15$$
$$\text{class4}: \quad 4,8,12$$

The sample program prints out the classes and ought to agree with this list.

```
PROGRAM d8r10(input,output);
(* driver for routine ECLASS *)
(*$I MODFILE.PAS *)
CONST
   m = 11;
   n = 15;
TYPE
   IntegerArrayNP = ARRAY [1..n] OF integer;
   IntegerArrayMP = ARRAY [1..m] OF integer;
VAR
   i,j,k,lclas,nclass: integer;
   lista,listb: IntegerArrayMP;
   nf,nflag,nsav: IntegerArrayNP;
(*$I ECLASS.PAS *)
BEGIN
   lista[1] := 1; lista[2] := 1; lista[3] := 5; lista[4] := 2;
   lista[5] := 6; lista[6] := 2; lista[7] := 7; lista[8] := 11;
   lista[9] := 3; lista[10] := 4; lista[11] := 12;
   listb[1] := 5; listb[2] := 9; listb[3] := 13; listb[4] := 6;
   listb[5] := 10; listb[6] := 14; listb[7] := 3; listb[8] := 7;
   listb[9] := 15; listb[10] := 8; listb[11] := 4;
   eclass(nf,n,lista,listb,m);
   FOR i := 1 TO n DO nflag[i] := 1;
   writeln;
   writeln('Numbers from 1-15 divided according to');
   writeln('their value modulo 4:');
   writeln;
   lclas := 0;
   FOR i := 1 TO n DO BEGIN
      nclass := nf[i];
```

```
      IF nflag[nclass] <> 0 THEN BEGIN
         nflag[nclass] := 0;
         lclas := lclas+1;
         k := 0;
         FOR j := i TO n DO BEGIN
            IF nf[j] = nf[i] THEN BEGIN
               k := k+1;
               nsav[k] := j
            END
         END;
         write('Class',lclas:2,':    ');
         FOR j := 1 TO k DO write(nsav[j]:3);
         writeln
      END
   END
END.
```

eclazz performs the same analysis but figures the equivalences from a boolean function equiv(i,j) that tells whether i and j are in the same equivalence class. In d8r11, equiv is defined as TRUE if (i MOD 4) and (j MOD 4) are the same. It is otherwise FALSE.

```
PROGRAM d8r11(input,output);
(* driver for routine ECLAZZ *)
(*$I MODFILE.PAS *)
CONST
   n = 15;
TYPE
   IntegerArrayNP = ARRAY [1..n] OF integer;
VAR
   i,j,k,lclas,nclass: integer;
   nf,nflag,nsav: IntegerArrayNP;
FUNCTION equiv(i,j: integer): boolean;
BEGIN
   IF i MOD 4 = j MOD 4 THEN
      equiv := TRUE
   ELSE
      equiv := FALSE
END;
(*$I ECLAZZ.PAS *)
BEGIN
   eclazz(nf,n);
   FOR i := 1 TO n DO nflag[i] := 1;
   writeln;
   writeln('Numbers from 1-15 divided according to');
   writeln('their value modulo 4:');
   lclas := 0;
   FOR i := 1 TO n DO BEGIN
      nclass := nf[i];
      IF nflag[nclass] <> 0 THEN BEGIN
         nflag[nclass] := 0;
         lclas := lclas+1;
         k := 0;
         FOR j := i TO n DO BEGIN
            IF nf[j] = nf[i] THEN BEGIN
               k := k+1;
               nsav[k] := j
```

```
            END
        END;
        write('Class',lclas:2,':    ');
        FOR j := 1 TO k DO write(nsav[j]:3);
        writeln
      END
    END
END.
```

Appendix

File tarray.dat:

```
Test data for chapter 8:
  29.82 71.51  3.30 87.44 53.42 63.16 89.10 25.75 93.16 27.72
  71.58 48.34 53.11 18.34 27.13 60.31 83.34 22.81 66.84 52.91
  53.42 15.22  8.01 53.39 76.12 79.09 67.61 38.39 24.81 73.21
  13.42 52.10 34.86 99.83 38.46 81.59 61.75 79.62 93.39  3.21
  99.34 92.22 94.29  7.03  6.67 89.35 83.14  9.01 12.68 62.22
   2.95 85.02 95.82 73.96 49.29 77.72 36.65  3.48 48.98 71.83
   1.41  9.48 32.37 89.95 28.39 79.36 54.05 46.08 11.67 37.78
  77.17 74.33 10.13  4.62 49.95 68.40 19.40 34.06  4.11 98.40
  42.44 64.14 89.41 52.99 71.79  3.94 19.73 44.91 71.44 59.10
  27.54 15.67 67.95 55.61 26.05 25.01 82.09 89.67 57.08 38.27
```

Chapter 9: Root Finding and Sets of Equations

Chapter 9 of Numerical Recipes deals primarily with the problem of finding roots to equations, and treats the problem in greatest detail in one dimension. We begin with a general-purpose routine called scrsho *that produces a crude graph of a given function on a specified interval. It is used for low-resolution plotting to investigate the properties of the function. With this in hand, we add bracketing routines* zbrac *and* zbrak. *The first of these takes a function and an interval, and expands the interval geometrically until it brackets a root. The second breaks the interval into N subintervals of equal size. It then reports any intervals that contain at least one root. Once bracketed, roots can be found by a number of other routines.* rtbis *finds such roots by bisection.* rtflsp *and* rtsec *use the method of false position and the secant method, respectively.* zbrent *uses a combination of methods to give assured and relatively efficient convergence.* rtnewt *implements the Newton-Raphson root finding method, while* rtsafe *combines it with bisection to correct for its risky global convergence properties.*

For finding the roots of polynomials, laguer *is handy, and when combined with its driver* zroots *it can find all roots of a polynomial having complex coefficients. When you have some tentative complex roots of a real polynomial, they can be polished by* qroot, *which employs Bairstow's method.*

In multiple dimensions, root-finding requires some foresight. However, if you can identify the neighborhood of a root of a system of nonlinear equations, then mnewt *will help you to zero in using Newton-Raphson.*

$$\star \quad \star \quad \star \quad \star$$

scrsho is a primitive graphing routine that will print graphs on virtually any terminal or printer. Sample program d9r1 demonstrates it by graphing the zero-order Bessel function J_0.

```
PROGRAM d9r1(input,output);
(* driver for routine SCRSHO *)
(*$I MODFILE.PAS *)
(*$I BESSJO.PAS *)
FUNCTION fx(x: real): real;
BEGIN
   fx := bessj0(x)
END;
(*$I SCRSHO.PAS *)
BEGIN
   scrsho
END.
```

110

zbrac is a root-bracketing routine that works by expanding the range of an interval geometrically until it brackets a root. Sample program d9r2 applies it to the Bessel function J_0. It starts with the ten intervals (1.0, 2.0), (2.0, 3.0), etc., and expands each until it contains a root. Then it prints the interval limits, and the function J_0 evaluated at these limits. The two values of J_0 should have opposite signs.

```pascal
PROGRAM d9r2(input,output);
(* driver for routine ZBRAC *)
(*$I MODFILE.PAS *)
VAR
    succes: boolean;
    i: integer;
    x1,x2: real;
(*$I BESSJO.PAS *)
FUNCTION fx(x: real): real;
BEGIN
    fx := bessj0(x)
END;
(*$I ZBRAC.PAS *)
BEGIN
    writeln('bracketing values:':21,'function values:':23);
    writeln('x1':6,'x2':10,'bessj0(x1)':21,'bessj0(x2)':12);
    FOR i := 1 TO 10 DO BEGIN
        x1 := i;
        x2 := x1+1.0;
        zbrac(x1,x2,succes);
        IF succes THEN
            writeln(x1:7:2,x2:10:2,' ':7,fx(x1):12:6,fx(x2):12:6)
    END
END.
```

zbrak is much like zbrac except that it takes an interval and subdivides it into N equal parts. It then identifies any of the subintervals that contain roots. Sample program d9r3 looks for roots of $J_0(x)$ between x1 = 1.0 and x2 = 50.0 by allowing zbrak to divide the interval into $N = 100$ parts. If there are no roots spaced closer than $\Delta x = 0.49$, then it will find brackets for all roots in this region. The limits of bracketing intervals, as well as function values at these limits, are printed, and again the function values at the end of each interval ought to be of opposite sign. There are 16 roots of J_0 between 1 and 50.

```pascal
PROGRAM d9r3(input,output);
(* driver for routine ZBRAK *)
(*$I MODFILE.PAS *)
CONST
    n = 100;
    nbmax = 20;
    x1 = 1.0;
    x2 = 50.0;
TYPE
    RealArrayNBMAX = ARRAY [1..nbmax] OF real;
VAR
    i,nb: integer;
    xb1,xb2: RealArrayNBMAX;
(*$I BESSJO.PAS *)
FUNCTION fx(x: real): real;
BEGIN
```

```
    fx := bessj0(x)
END;
(*$I ZBRAK.PAS *)
BEGIN
    nb := nbmax;
    zbrak(x1,x2,n,xb1,xb2,nb);
    writeln;
    writeln('brackets for roots of bessj0:');
    writeln('lower':22,'upper':10,'f(lower)':16,'f(upper)':10);
    FOR i := 1 TO nb DO
        writeln('  root ',i:2,' ':4,xb1[i]:10:4,xb2[i]:10:4,
            ' ':4,fx(xb1[i]):10:4,fx(xb2[i]):10:4)
END.
```

Routine rtbis begins with the brackets for a root and finds the root itself by bisection, The accuracy with which the root is found is determined by parameter xacc. Sample program d9r4 finds all the roots of Bessel function $J_0(x)$ between $x1 = 1.0$ and $x2 = 50.0$. In this case xacc is specified to be about 10^{-6} of the value of the root itself (actually, 10^{-6} of the center of the interval being bisected). The roots root are listed, as well as $J_0(\text{root})$ to verify their accuracy.

```
PROGRAM d9r4(input,output);
(* driver for routine RTBIS *)
(*$I MODFILE.PAS *)
CONST
    n = 100;
    nbmax = 20;
    x1 = 1.0;
    x2 = 50.0;
TYPE
    RealArrayNBMAX = ARRAY [1..nbmax] OF real;
VAR
    i,nb: integer;
    xacc,root: real;
    xb1,xb2: RealArrayNBMAX;
(*$I BESSJ0.PAS *)
FUNCTION fx(x: real): real;
BEGIN
    fx := bessj0(x)
END;
(*$I ZBRAK.PAS *)
(*$I RTBIS.PAS *)
BEGIN
    nb := nbmax;
    zbrak(x1,x2,n,xb1,xb2,nb);
    writeln;
    writeln('roots of bessj0:');
    writeln('x':20,'f(x)':15);
    FOR i := 1 TO nb DO BEGIN
        xacc := (1.0e-6)*(xb1[i]+xb2[i])/2.0;
        root := rtbis(xb1[i],xb2[i],xacc);
        writeln('  root ',i:2,' ',root:12:6,fx(root):14:6)
    END
END.
```

The next five sample programs, d9r5–d9r9, are essentially identical to the one just discussed, except for the root-finder they employ. **d9r5** calls **rtflsp**, finding the root by "false position". **d9r6** calls **rtsec** and uses the secant method. **d9r7** uses **zbrent** to give reliable and efficient convergence. The Newton-Raphson method implemented in **rtnewt** is demonstrated by **d9r8**, and **d9r9** calls **rtsafe**, which improves upon **rtnewt** by combining it with bisection to achieve better global convergence. The latter two programs include a procedure **funcd** that returns the value of the function and its derivative at a given x. In the case of test function $J_0(x)$ the derivative is $-J_1(x)$, and is conveniently in our collection of special functions.

```
PROGRAM d9r5(input,output);
(* driver for routine RTFLSP *)
(*$I MODFILE.PAS *)
CONST
    n = 100;
    nbmax = 20;
    x1 = 1.0;
    x2 = 50.0;
TYPE
    RealArrayNBMAX = ARRAY [1..nbmax] OF real;
VAR
    i,nb: integer;
    root,xacc: real;
    xb1,xb2: RealArrayNBMAX;
(*$I BESSJO.PAS *)
FUNCTION fx(x: real): real;
BEGIN
    fx := bessj0(x)
END;
(*$I ZBRAK.PAS *)
(*$I RTFLSP.PAS *)
BEGIN
    nb := nbmax;
    zbrak(x1,x2,n,xb1,xb2,nb);
    writeln;
    writeln('roots of bessj0:');
    writeln('x':20,'f(x)':15);
    FOR i := 1 TO nb DO BEGIN
        xacc := (1.0e-6)*(xb1[i]+xb2[i])/2.0;
        root := rtflsp(xb1[i],xb2[i],xacc);
        writeln('  root ',i:2,'  ',root:12:6,fx(root):14:6)
    END
END.

PROGRAM d9r6(input,output);
(* driver for routine RTSEC *)
(*$I MODFILE.PAS *)
CONST
    n = 100;
    nbmax = 20;
    x1 = 1.0;
    x2 = 50.0;
TYPE
    RealArrayNBMAX = ARRAY [1..nbmax] OF real;
VAR
    i,nb: integer;
    root,xacc: real;
```

```
   xb1,xb2: RealArrayNBMAX;
(*$I BESSJO.PAS *)
FUNCTION fx(x: real): real;
BEGIN
   fx := bessj0(x)
END;
(*$I ZBRAK.PAS *)
(*$I RTSEC.PAS *)
BEGIN
   nb := nbmax;
   zbrak(x1,x2,n,xb1,xb2,nb);
   writeln;
   writeln('roots of bessj0:');
   writeln('x':20,'f(x)':15);
   FOR i := 1 TO nb DO BEGIN
      xacc := (1.0e-6)*(xb1[i]+xb2[i])/2.0;
      root := rtsec(xb1[i],xb2[i],xacc);
      writeln('  root ',i:2,'  ',root:12:6,fx(root):14:6)
   END
END.


PROGRAM d9r7(input,output);
(* driver for routine ZBRENT *)
(*$I MODFILE.PAS *)
CONST
   n = 100;
   nbmax = 20;
   x1 = 1.0;
   x2 = 50.0;
TYPE
   RealArrayNBMAX = ARRAY [1..nbmax] OF real;
VAR
   i,nb: integer;
   root,tol: real;
   xb1,xb2: RealArrayNBMAX;
(*$I BESSJO.PAS *)
FUNCTION fx(x: real): real;
BEGIN
   fx := bessj0(x)
END;
(*$I ZBRAK.PAS *)
(*$I ZBRENT.PAS *)
BEGIN
   nb := nbmax;
   zbrak(x1,x2,n,xb1,xb2,nb);
   writeln;
   writeln('Roots of bessj0:');
   writeln('x':18,'f(x)':15);
   FOR i := 1 TO nb DO BEGIN
      tol := (1.0e-6)*(xb1[i]+xb2[i])/2.0;
      root := zbrent(xb1[i],xb2[i],tol);
      writeln('root ',i:2,'  ',root:12:6,fx(root):14:6)
   END
END.
```

```
PROGRAM d9r8(input,output);
(* driver for routine RTNEWT *)
(*$I MODFILE.PAS *)
CONST
   n = 100;
   nbmax = 20;
   x1 = 1.0;
   x2 = 50.0;
TYPE
   RealArrayNBMAX = ARRAY [1..nbmax] OF real;
VAR
   i,nb: integer;
   root,xacc: real;
   xb1,xb2: RealArrayNBMAX;
(*$I BESSJO.PAS *)
(*$I BESSJ1.PAS *)
FUNCTION fx(x: real): real;
BEGIN
   fx := bessj0(x)
END;
PROCEDURE funcd(x: real; VAR fn,df: real);
BEGIN
   fn := fx(x);
   df := -bessj1(x)
END;
(*$I ZBRAK.PAS *)
(*$I RTNEWT.PAS *)
BEGIN
   nb := nbmax;
   zbrak(x1,x2,n,xb1,xb2,nb);
   writeln;
   writeln('roots of bessj0:');
   writeln('x':20,'f(x)':15);
   FOR i := 1 TO nb DO BEGIN
      xacc := (1.0e-6)*(xb1[i]+xb2[i])/2.0;
      root := rtnewt(xb1[i],xb2[i],xacc);
      writeln('  root ',i:2,'  ',root:12:6,fx(root):14:6)
   END
END.

PROGRAM d9r9(input,output);
(* driver for routine RTSAFE *)
(*$I MODFILE.PAS *)
CONST
   n = 100;
   nbmax = 20;
   x1 = 1.0;
   x2 = 50.0;
TYPE
   RealArrayNBMAX = ARRAY [1..nbmax] OF real;
VAR
   i,nb: integer;
   root,xacc: real;
   xb1,xb2: RealArrayNBMAX;
(*$I BESSJO.PAS *)
(*$I BESSJ1.PAS *)
FUNCTION fx(x: real): real;
```

```
BEGIN
   fx := bessj0(x)
END;
PROCEDURE funcd(x: real; VAR fn,df: real);
BEGIN
   fn := fx(x);
   df := -bessj1(x)
END;
(*$I ZBRAK.PAS *)
(*$I RTSAFE.PAS *)
BEGIN
   nb := nbmax;
   zbrak(x1,x2,n,xb1,xb2,nb);
   writeln;
   writeln('roots of bessj0:');
   writeln('x':19,'f(x)':16);
   FOR i := 1 TO nb DO BEGIN
      xacc := (1.0e-6)*(xb1[i]+xb2[i])/2.0;
      root := rtsafe(xb1[i],xb2[i],xacc);
      writeln(' root ',i:2,'  ',root:12:6,fx(root):14:6)
   END
END.
```

Routine `laguer` finds the roots of a polynomial with complex coefficients. The polynomial of degree M is specified by $M + 1$ coefficients which, in sample program d9r10, are specified in the complex array a. The polynomial in this case is

$$F(x) = x^4 - (1 + 2i)x^2 + 2i$$

The four roots of this polynomial are $x = 1.0$, $x = -1.0$, $x = 1 + i$, and $x = -(1 + i)$. `laguer` proceeds on the basis of a trial root, and attempts to converge to true roots. The root to which it converges depends on the trial value. The program tries a series of complex trial values along the line in the imaginary plane from $-1.0 - i$ to $1.0 + i$. The actual roots to which it converges are compared to all previously found values, and if different, are printed.

```
PROGRAM d9r10(input,output);
(* driver for routine LAGUER *)
(*$I MODFILE.PAS *)
CONST
   m = 4;
   mp1 = 5;        (* mp1=m+1 *)
   ntry = 21;
   eps = 1.0e-6;
TYPE
   Complex = RECORD
                 r,i: real
             END;
   ComplexArrayMp1 = ARRAY [1..mp1] OF Complex;
   ComplexArrayNTRY = ARRAY [1..ntry] OF Complex;
VAR
   i,iflag,j,n: integer;
   polish: boolean;
   x: Complex;
   a: ComplexArrayMp1;
   y: ComplexArrayNTRY;
```

```
(*$I LAGUER.PAS *)
BEGIN
   a[1].r := 0.0; a[1].i := 2.0;
   a[2].r := 0.0; a[2].i := 0.0;
   a[3].r := -1.0; a[3].i := -2.0;
   a[4].r := 0.0; a[4].i := 0.0;
   a[5].r := 1.0; a[5].i := 0.0;
   writeln;
   writeln('Roots of polynomial x^4-(1+2i)*x^2+2i');
   writeln('real':15,'complex':13);
   n := 0;
   polish := false;
   FOR i := 1 TO ntry DO BEGIN
      x.r := (i-11.0)/10.0;
      x.i := (i-11.0)/10.0;
      laguer(a,m,x,eps,polish);
      IF n = 0 THEN BEGIN
         n := 1;
         y[1] := x;
         writeln(n:5,x.r:12:6,x.i:12:6)
      END ELSE BEGIN
         iflag := 0;
         FOR j := 1 TO n DO
            IF sqrt(sqr(x.r-y[j].r)+sqr(x.i-y[j].i))
               <= eps*sqrt(sqr(x.r)+sqr(x.i)) THEN
               iflag := 1;
         IF iflag = 0 THEN BEGIN
            n := n+1;
            y[n] := x;
            writeln(n:5,x.r:12:6,x.i:12:6)
         END
      END
   END
END.
```

zroots is a driver for laguer. Sample program d9r11 exercises zroots, finding all four roots of the same polynomial as the previous routine.

```
PROGRAM d9r11(input,output);
(* driver for routine ZROOTS *)
(*$I MODFILE.PAS *)
CONST
   m = 4;
   mp1 = 5;          (* mp1=m+1 *)
TYPE
   Complex = RECORD
               r,i: real
             END;
   ComplexArrayMp1 = ARRAY [1..mp1] OF Complex;
VAR
   i: integer;
   polish: boolean;
   a,roots: ComplexArrayMp1;
(*$I LAGUER.PAS *)
(*$I ZROOTS.PAS *)
BEGIN
   a[1].r := 0.0; a[1].i := 2.0;
```

```
   a[2].r := 0.0; a[2].i := 0.0;
   a[3].r := -1.0; a[3].i := -2.0;
   a[4].r := 0.0; a[4].i := 0.0;
   a[5].r := 1.0; a[5].i := 0.0;
   writeln('Roots of the polynomial x^4-(1+2i)*x^2+2i');
   polish := true;
   zroots(a,m,roots,polish);
   writeln;
   writeln('root #':14,'real':13,'imag.':13);
   FOR i := 1 TO m DO
      writeln(i:11,' ':5,roots[i].r:12:6,roots[i].i:12:6);
END.
```

qroot is used for finding quadratic factors of polynomials with real coefficients. In the case of sample program d9r12, the polynomial is

$$P(x) = x^6 - 6x^5 + 16x^4 - 24x^3 + 25x^2 - 18x + 10.$$

The program proceeds like the driver for laguer. Successive trial values for quadratic factors $x^2 + Bx + C$ (in the form of guesses for B and C) are made, and for each trial, qroot converges on correct values. If the B and C which are found are unlike any previous values, then they are printed. By this means, all three quadratic factors are located. You can, of course, compare their product to the polynomial above.

```
PROGRAM d9r12(input,output);
(* driver for routine QROOT *)
(*$I MODFILE.PAS *)
CONST
   n = 7;
   nv = 3;
   eps = 1.0e-6;
   ntry = 10;
   tiny = 1.0e-5;
TYPE
   RealArrayNP = ARRAY [1..n] OF real;
   RealArrayNV = ARRAY [1..nv] OF real;
   RealArrayNTRY = ARRAY [1..ntry] OF real;
VAR
   i,j,nflag,nroot: integer;
   p: RealArrayNP;
   b,c: RealArrayNTRY;
(*$I POLDIV.PAS *)
(*$I QROOT.PAS *)
BEGIN
   p[1] := 10.0; p[2] := -18.0; p[3] := 25.0; p[4] := -24.0;
   p[5] := 16.0; p[6] := -6.0; p[7] := 1.0;
   writeln;
   writeln('P(x) := x^6-6x^5+16x^4-24x^3+25x^2-18x+10');
   writeln('Quadratic factors x^2+bx+c');
   writeln;
   writeln('factor':6,'b':10,'c':12);
   writeln;
   nroot := 0;
   FOR i := 1 TO ntry DO BEGIN
      c[i] := 0.5*i;
      b[i] := -0.5*i;
      qroot(p,n,b[i],c[i],eps);
```

```
      IF nroot = 0 THEN BEGIN
        writeln(nroot:4,'    ',b[i]:12:6,c[i]:12:6);
        nroot := 1
      END ELSE BEGIN
        nflag := 0;
        FOR j := 1 TO nroot DO
          IF (abs(b[i]-b[j]) < tiny)
            AND (abs(c[i]-c[j]) < tiny) THEN nflag := 1;
          IF nflag = 0 THEN BEGIN
            writeln(nroot:4,'    ',b[i]:12:6,c[i]:12:6);
            nroot := nroot+1
          END
        END
      END
    END.
```

Finally, mnewt looks for roots of multiple nonlinear equations. In order to run a sample program d9r13 we supply a procedure usrfun that returns the matrix alpha of partial derivatives of the functions with respect to each of the variables, and vector beta, containing the negatives of the function values. The sample program tries to find sets of variables that solve the four equations

$$-x_1^2 - x_2^2 - x_3^2 + x_4 = 0$$
$$x_1^2 + x_2^2 + x_3^2 + x_4^2 - 1 = 0$$
$$x_1 - x_2 = 0$$
$$x_2 - x_3 = 0$$

You will probably be able to find the two solutions to this set even without mnewt, noting that $x_1 = x_2$ and $x_2 = x_3$. If not, simply take the output from mnewt and plug it into these equations for verification. The output from mnewt should convince you of the need for good starting values.

```
PROGRAM d9r13(input,output);
(* driver for routine MNEWT *)
(*$I MODFILE.PAS *)
CONST
   ntrial = 5;
   tolx = 1.0e-6;
   n = 4;
   np = n;
   tolf = 1.0e-6;
TYPE
   RealArrayNP = ARRAY [1..n] OF real;
   RealArrayNPbyNP = ARRAY [1..n,1..n] OF real;
   IntegerArrayNP = ARRAY [1..n] OF integer;
VAR
   i,j,k,kk,l: integer;
   xx: real;
   x,beta: RealArrayNP;
   alpha: RealArrayNPbyNP;
PROCEDURE usrfun(VAR x: RealArrayNP; n: integer; VAR alpha: RealArrayNPbyNP;
      VAR beta: RealArrayNP);
(* Programs using routine USRFUN must define the types
TYPE
   RealArrayNP = ARRAY [1..n] OF real;
```

```
     RealArrayNPbyNP = ARRAY [1..n,1..n] OF real;
in the main routine. *)
BEGIN
   alpha[1,1] := 2.0*x[1];
   alpha[1,2] := 2.0*x[2];
   alpha[1,3] := 2.0*x[3];
   alpha[1,4] := -3.0;
   alpha[2,1] := 2.0*x[1];
   alpha[2,2] := 2.0*x[2];
   alpha[2,3] := 2.0*x[3];
   alpha[2,4] := -2.0*x[4];
   alpha[3,1] := 1.0;
   alpha[3,2] := -1.0;
   alpha[3,3] := 0.0;
   alpha[3,4] := 0.0;
   alpha[4,1] := 0.0;
   alpha[4,2] := 1.0;
   alpha[4,3] := -1.0;
   alpha[4,4] := 0.0;
   beta[1] := -sqr(x[1])-sqr(x[2])-sqr(x[3])+3.0*x[4];
   beta[2] := -sqr(x[1])-sqr(x[2])-sqr(x[3])+sqr(x[4])+1.0;
   beta[3] := -x[1]+x[2];
   beta[4] := -x[2]+x[3]
END;
(*$I LUBKSB.PAS *)
(*$I LUDCMP.PAS *)
(*$I MNEWT.PAS *)
BEGIN
   FOR l := 0 TO 1 DO BEGIN
      kk := 2*l-1;
      FOR k := 1 TO 3 DO BEGIN
         xx := 0.2*k*kk;
         writeln('Starting vector number',k:2);
         FOR i := 1 TO 4 DO
            x[i] := xx+0.2*i;
         FOR i := 1 TO 4 DO
            writeln('x[':7,i:1,'] := ',x[i]:5:2);
         writeln;
         FOR j := 1 TO ntrial DO BEGIN
            mnewt(1,x,n,tolx,tolf);
            usrfun(x,n,alpha,beta);
            writeln('i':5,'x[i]':13,'f':13);
            FOR i := 1 TO n DO
               writeln(i:5,'    ',x[i]:12,'    ',-beta[i]:12);
            writeln;
            writeln('press RETURN to continue...');
            readln
         END
      END
   END
END.
```

Chapter 10: Minimization and Maximization of Functions

Chapter 10 of Numerical Recipes deals with finding the maxima and minima of functions. The task has two parts, first the discovery of one or more bracketing intervals, and then the convergence to an extremum. mnbrak *begins with two specified abscissas of a function and searches in the "downhill" direction for brackets of a minimum.* golden *can then take a bracketing triplet and perform a golden section search to a specified precision, for the minimum itself. When you are not concerned with worst-case examples, but only very efficient average-case performance, Brent's method (routine* brent*) is recommended. In the event that means are at hand for calculating the function's derivative as well as its value, consider* dbrent.

Multidimensional minimization strategies are generally based on the one-dimensional algorithms. Our single example of an algorithm that is not so based is amoeba*, which utilizes the downhill simplex method. Among the ones that do use one-dimensional methods are* powell*,* frprmn*, and* dfpmin*. These three all make calls to* linmin*, a procedure that minimizes a function along a given direction in space.* linmin *in turn uses the one-dimensional algorithm* brent*, if derivatives are not known, or* dbrent *if they are.* powell *uses only function values and minimizes along an artfully chosen set of favorable directions.* frprmn *uses a Fletcher-Reeves-Polak-Ribiere minimization and requires the calculation of derivatives for the function.* dfpmin *uses a variant of the Davidon-Fletcher-Powell variable metric method. This, too, requires calculation of derivatives.*

The chapter ends with two topics of somewhat different nature. The first is linear programming, which deals with the maximation of a linear combination of variables, subject to linear constraints. This problem is dealt with by the simplex method in routine simplx*. The second is the subject of large scale optimization, which is illustrated with the method of simulated annealing, and applied particularly to the "travelling salesman" problem in routine* anneal*.*

⋆ ⋆ ⋆ ⋆

mnbrak searches a given function for a minimum. Given two values ax and bx of abscissa, it searches in the downward direction until it can find three new values ax,bx,cx that bracket a minimum. fa,fb,fc are the values of the function at these points. Sample program d10r1 is a simple application of mnbrak applied to the Bessel function J_0. It tries a series of starting values ax,bx each encompassing an interval of length 1.0. mnbrak then finds several bracketing intervals of various minima of J_0.

```pascal
PROGRAM d10r1(input,output);
(* driver for routine MNBRAK *)
(*$I MODFILE.PAS *)
VAR
    ax,bx,cx,fa,fb,fc: real;
    i: integer;
(*$I BESSJ0.PAS *)
FUNCTION func(x: real): real;
BEGIN
    func := bessj0(x)
END;
(*$I MNBRAK.PAS *)
BEGIN
    FOR i := 1 TO 10 DO BEGIN
        ax := i*0.5;
        bx := (i+1.0)*0.5;
        mnbrak(ax,bx,cx,fa,fb,fc);
        writeln('a':14,'b':12,'c':12);
        writeln('x':3,'  ',ax:12:6,bx:12:6,cx:12:6);
        writeln('f':3,'  ',fa:12:6,fb:12:6,fc:12:6)
    END
END.
```

Routine golden continues the minimization process by taking a bracketing triplet ax,bx,cx and performing a golden section search to isolate the contained minimum to a stated precision tol. Sample program d10r2 again uses J_0 as the test function. Using intervals (ax,bx) of length 1.0 it uses mnbrak to bracket all minima between $x = 0.0$ and $x = 100.0$. Some minima are bracketed more than once. On each pass, the bracketed solution is tracked down by golden. It is then compared to all previously located minima, and if different it is added to the collection by incrementing nmin (number of minima found) and adding the location xmin of the minima to the list in array amin. As a check of golden, the routine prints out the value of J_0 at the minimum, and also the value of J_1, which ought to be zero at extrema of J_0.

```pascal
PROGRAM d10r2(input,output);
(* driver for routine GOLDEN *)
(*$I MODFILE.PAS *)
CONST
    tol = 1.0e-6;
    eql = 1.0e-3;
TYPE
    RealArray20 = ARRAY [1..20] OF real;
VAR
    ax,bx,cx,fa,fb,fc,xmin,gold: real;
    i,iflag,j,nmin: integer;
    amin: RealArray20;
(*$I BESSJ0.PAS *)
(*$I BESSJ1.PAS *)
FUNCTION func(x: real): real;
BEGIN
    func := bessj0(x)
END;
```

```
(*$I MNBRAK.PAS *)
(*$I GOLDEN.PAS *)
BEGIN
   nmin := 0;
   writeln('minima of the function bessj0');
   writeln('min. #':10,'x':8,'bessj0(x)':17,'bessj1(x)':12);
   FOR i := 1 TO 100 DO BEGIN
      ax := i;
      bx := i+1.0;
      mnbrak(ax,bx,cx,fa,fb,fc);
      gold := golden(ax,bx,cx,tol,xmin);
      IF nmin = 0 THEN BEGIN
         amin[1] := xmin;
         nmin := 1;
         writeln(nmin:7,xmin:15:6,bessj0(xmin):12:6,
            bessj1(xmin):12:6)
      END ELSE BEGIN
         iflag := 0;
         FOR j := 1 TO nmin DO BEGIN
            IF abs(xmin-amin[j]) <= eql*xmin THEN
               iflag := 1;
         END;
         IF iflag = 0 THEN BEGIN
            nmin := nmin+1;
            amin[nmin] := xmin;
            writeln(nmin:7,xmin:15:6,bessj0(xmin):12:6,
               bessj1(xmin):12:6)
         END
      END
   END
END.
```

There are two other routines presented which also take the bracketing triplet ax,bx,cx from mnbrak and find the contained minimum. They are brent and dbrent. The sample programs for these two, d10r3 and d10r4, are virtually identical to that used on golden. Note that dbrent is only used when the derivative, in this case (−bessj1), can be calculated conveniently.

```
PROGRAM d10r3(input,output);
(* driver for routine BRENT *)
(*$I MODFILE.PAS *)
CONST
   tol = 1.0e-6;
   eql = 1.0e-4;
TYPE
   RealArray20 = ARRAY [1..20] OF real;
VAR
   ax,bx,cx,fa,fb,fc,xmin,bren: real;
   i,iflag,j,nmin: integer;
   amin: RealArray20;
(*$I BESSJ0.PAS *)
(*$I BESSJ1.PAS *)
FUNCTION func(x: real): real;
BEGIN
   func := bessj0(x)
END;
(*$I MNBRAK.PAS *)
```

```
(*$I BRENT.PAS *)
BEGIN
   nmin := 0;
   writeln;
   writeln('minima of the function bessj0');
   writeln('min. #':10,'x':8,'bessj0(x)':17,'bessj1(x)':12);
   FOR i := 1 TO 100 DO BEGIN
      ax := i;
      bx := i+1.0;
      mnbrak(ax,bx,cx,fa,fb,fc);
      bren := brent(ax,bx,cx,tol,xmin);
      IF nmin = 0 THEN BEGIN
         amin[1] := xmin;
         nmin := 1;
         writeln(nmin:7,xmin:15:6,bessj0(xmin):12:6,
            bessj1(xmin):12:6)
      END ELSE BEGIN
         iflag := 0;
         FOR j := 1 TO nmin DO
            IF abs(xmin-amin[j]) <= eql*xmin THEN
               iflag := 1;
         IF iflag = 0 THEN BEGIN
            nmin := nmin+1;
            amin[nmin] := xmin;
            writeln(nmin:7,xmin:15:6,bessj0(xmin):12:6,
               bessj1(xmin):12:6)
         END
      END
   END
END.

PROGRAM d10r4(input,output);
(* driver for routine DBRENT *)
(*$I MODFILE.PAS *)
CONST
   tol = 1.0e-6;
   eql = 1.0e-4;
TYPE
   RealArray20 = ARRAY [1..20] OF real;
VAR
   ax,bx,cx,fa,fb,fc,xmin,dbr: real;
   i,iflag,j,nmin: integer;
   amin: RealArray20;
(*$I BESSJ1.PAS *)
FUNCTION dfunc(x: real): real;
BEGIN
   dfunc := -bessj1(x)
END;
(*$I BESSJ0.PAS *)
FUNCTION func(x: real): real;
BEGIN
   func := bessj0(x)
END;
(*$I MNBRAK.PAS *)
(*$I DBRENT.PAS *)
BEGIN
   nmin := 0;
```

```
writeln;
writeln('minima of the function bessj0');
writeln('min. #':10,'x':8,
    'bessj0(x)':16,'bessj1(x)':12,'DBRENT':11);
FOR i := 1 TO 100 DO BEGIN
    ax := i;
    bx := i+1.0;
    mnbrak(ax,bx,cx,fa,fb,fc);
    dbr := dbrent(ax,bx,cx,tol,xmin);
    IF nmin = 0 THEN BEGIN
        amin[1] := xmin;
        nmin := 1;
        writeln(nmin:7,xmin:15:6,func(xmin):12:6,
            dfunc(xmin):12:6,dbr:12:6);
    END ELSE BEGIN
        iflag := 0;
        FOR j := 1 TO nmin DO
            IF abs(xmin-amin[j]) <= eql*xmin THEN
                iflag := 1;
        IF iflag = 0 THEN BEGIN
            nmin := nmin+1;
            amin[nmin] := xmin;
            writeln(nmin:7,xmin:15:6,func(xmin):12:6,
                dfunc(xmin):12:6,dbr:12:6)
        END
    END
END
END.
```

Numerical Recipes presents several methods for minimization in multiple dimensions. Among these, the downhill simplex method carried out by amoeba is the only
one that does not treat the problem as a series of one-dimensional minimizations.
As input, amoeba requires the coordinates of $N + 1$ vertices of a starting simplex
in N-dimensional space, and the values y of the function at each of these vertices.
Sample program d10r5 tries the method out on the exotic function

$$\text{func} = 0.6 - J_0[(x - 0.5)^2 + (y - 0.6)^2 + (z - 0.7)^2]$$

which has a minimum at $(x, y, z) = (0.5, 0.6, 0.7)$. As vertices of the starting simplex,
specified by the array p, we use $(0, 0, 0)$, $(1, 0, 0)$, $(0, 1, 0)$, and $(0, 0, 1)$. A vector x[i]
is set successively to each vertex to allow the evaluation of function values y. This
data is submitted to amoeba along with ftol=1.0e-6 to specify the tolerance on the
function value. The vertices and corresponding function values of the final simplex
are printed out, and you can easily check whether the specified tolerance is met.

```
PROGRAM d10r5(input,output);
(* driver for routine AMOEBA *)
(*$I MODFILE.PAS *)
CONST
    np = 3;
    mp = 4;
    ftol = 1.0e-6;
TYPE
    RealArrayMPbyNP = ARRAY [1..mp,1..np] OF real;
    RealArrayMP = ARRAY [1..mp] OF real;
    RealArrayNP = ARRAY [1..np] OF real;
```

```
VAR
    i,nfunc,j,ndim: integer;
    x: RealArrayNP;
    y: RealArrayMP;
    p: RealArrayMPbyNP;
(*$I BESSJ0.PAS *)
FUNCTION func(VAR x: RealArrayNP): real;
(* calling function must define type
TYPE
    RealArrayNP = ARRAY [1..np] OF real;
where np is the physical dimension of the argument x. *)
BEGIN
    func := 0.6-bessj0(sqr(x[1]-0.5)+sqr(x[2]-0.6)+sqr(x[3]-0.7))
END;
(*$I AMOEBA.PAS *)
BEGIN
    p[1,1] := 0.0; p[1,2] := 0.0; p[1,3] := 0.0;
    p[2,1] := 1.0; p[2,2] := 0.0; p[2,3] := 0.0;
    p[3,1] := 0.0; p[3,2] := 1.0; p[3,3] := 0.0;
    p[4,1] := 0.0; p[4,2] := 0.0; p[4,3] := 1.0;
    ndim := np;
    FOR i := 1 TO mp DO BEGIN
        FOR j := 1 TO np DO x[j] := p[i,j];
        y[i] := func(x)
    END;
    amoeba(p,y,ndim,ftol,nfunc);
    writeln;
    writeln('Function evaluations: ',nfunc:3);
    writeln('Vertices of final 3-d simplex and');
    writeln('function values at the vertices:');
    writeln;
    writeln('i':3,
        'x[i]':10,'y[i]':12,'z[i]':12,'function':14);
    writeln;
    FOR i := 1 TO mp DO BEGIN
        write(i:3);
        FOR j := 1 TO np DO write(p[i,j]:12:6);
        writeln(y[i]:12:6)
    END;
    writeln;
    writeln('True minimum is at (0.5,0.6,0.7)')
END.
```

powell carries out one-dimensional minimizations along favorable directions in N-dimensional space. The function minimized must be called fnc, and in sample program d10r6 a function procedure is defined for

$$fnc(x,y,z) = \tfrac{1}{2} - J_0[(x-1)^2 + (y-2)^2 + (z-3)^2].$$

The program provides powell with a starting point P of $(3/2, 3/2, 5/2)$ and a set of initial directions, here chosen to be the unit directions $(1,0,0)$, $(0,1,0)$, and $(0,0,1)$. powell performs its one-dimensional minimizations with linmin, which is discussed next.

```
PROGRAM d10r6(input,output);
(* driver for routine POWELL *)
(*$I MODFILE.PAS *)
CONST
   ndim = 3;
   ftol = 1.0e-6;
TYPE
   RealArrayNP = ARRAY [1..ndim] OF real;
   RealArrayNPbyNP = ARRAY [1..ndim,1..ndim] OF real;
   RealArray3 = RealArrayNP;
VAR
   LinminNcom: integer;
   LinminPcom,LinminXicom: RealArrayNP;
   fret: real;
   i,iter,j: integer;
   p: RealArrayNP;
   xi: RealArrayNPbyNP;
(*$I BESSJ0.PAS *)
FUNCTION fnc(VAR x: RealArray3): real;
(* Programs using FNC must define the type
TYPE
   RealArray3 = ARRAY [1..3] OF real;
in the main routine. *)
BEGIN
   fnc := 0.5-bessj0(sqr(x[1]-1.0)+sqr(x[2]-2.0)+sqr(x[3]-3.0))
END;
(*$I F1DIM.PAS *)
(*$I MNBRAK.PAS *)
(*$I BRENT.PAS *)
(*$I LINMIN.PAS *)
(*$I POWELL.PAS *)
BEGIN
   xi[1,1] := 1.0; xi[1,2] := 0.0; xi[1,3] := 0.0;
   xi[2,1] := 0.0; xi[2,2] := 1.0; xi[2,3] := 0.0;
   xi[3,1] := 0.0; xi[3,2] := 0.0; xi[3,3] := 1.0;
   p[1] := 1.5; p[2] := 1.5; p[3] := 2.5;
   powell(p,xi,ndim,ftol,iter,fret);
   writeln('Iterations:',iter:3);
   writeln;
   writeln('Minimum found at: ');
   FOR i := 1 TO ndim DO write(p[i]:12:6);
   writeln;
   writeln;
   writeln('Minimum function value =',fret:12:6);
   writeln;
   writeln('True minimum of function is at:');
   writeln(1.0:12:6,2.0:12:6,3.0:12:6)
END.
```

linmin, as we have said, finds the minimum of a function along a direction in N-dimensional space. To use it we specify a point P and a direction vector xi, both in N-space. linmin then does the book-keeping required to treat the function as a function of position along this line, and minimizes the function with a conventional one-dimensional minimization routine. Sample program d10r7 feeds linmin the function

$$fnc(x,y,z) = (x-1)^2 + (y-1)^2 + (z-1)^2$$

which has a minimum at $(x, y, z) = (1, 1, 1)$. It also chooses point P to be the origin $(0, 0, 0)$, and tries a series of directions

$$\left(\sqrt{2}\cos\left(\frac{\pi}{2}\frac{i}{10.0}\right), \quad \sqrt{2}\sin\left(\frac{\pi}{2}\frac{i}{10.0}\right), \quad 1.0\right) \qquad i = 1, \ldots, 10$$

For each pass, the location of the minimum, and the value of the function at the minimum, are printed. Among the directions searched is the direction $(1, 1, 1)$. Along this direction, of course, the minimum function value should be zero and should occur at $(1, 1, 1)$.

```
PROGRAM d10r7(input,output);
(* driver for routine LINMIN *)
(*$I MODFILE.PAS *)
CONST
   ndim = 3;
   pio2 = 1.5707963;
TYPE
   RealArrayNP = ARRAY [1..ndim] OF real;
VAR
   LinminNcom: integer;
   LinminPcom,LinminXicom: RealArrayNP;
   fret,sr2,x: real;
   i,j: integer;
   p,xi: RealArrayNP;
FUNCTION fnc(VAR x: RealArrayNP): real;
(* calling routine must define type
TYPE
   RealArrayNP = ARRAY [1..np] OF real;
where np is the dimension of vector x. *)
VAR
   i: integer;
   f: real;
BEGIN
   f := 0.0;
   FOR i := 1 TO 3 DO f := f+sqr(x[i]-1.0);
   fnc := f
END;
(*$I F1DIM.PAS *)
(*$I MNBRAK.PAS *)
(*$I BRENT.PAS *)
(*$I LINMIN.PAS *)
BEGIN
   writeln;
   writeln('Minimum of a 3-d quadratic centered');
   writeln('at (1.0,1.0,1.0). Minimum is found');
   writeln('along a series of radials.');
   writeln;
   writeln('x':9,'y':12,'z':12,'minimum':14);
   FOR i := 0 TO 10 DO BEGIN
      x := pio2*i/10.0;
      sr2 := sqrt(2.0);
      xi[1] := sr2*cos(x);
      xi[2] := sr2*sin(x);
      xi[3] := 1.0;
      p[1] := 0.0;
      p[2] := 0.0;
```

```
         p[3] := 0.0;
         linmin(p,xi,ndim,fret);
         FOR j := 1 TO 3 DO write(p[j]:12:6);
         writeln(fret:12:6)
      END
END.
```

f1dim accompanies linmin and is the routine that makes an N-dimensional function effectively a one-dimensional function along a given line in N-space. There is little to check here, and our perfunctory demonstration of its use, in sample program d10r8, simply plots f1dim as a one dimensional function, given the function

$$\text{fnc}(x, y, z) = (x - 1)^2 + (y - 1)^2 + (z - 1)^2.$$

You get to choose the direction; then scrsho plots the function along this direction. Try the direction $(1, 1, 1)$ along which you should find a minimum value of fnc=0 at position $(1, 1, 1)$.

```
PROGRAM d10r8(input,output);
(* driver for routine F1DIM *)
(*$I MODFILE.PAS *)
CONST
   ndim = 3;
TYPE
   RealArrayNP = ARRAY [1..ndim] OF real;
VAR
   i,j,LinminNcom: integer;
   p,xi: RealArrayNP;
   LinminPcom,LinminXicom: RealArrayNP;
FUNCTION fnc(x: RealArrayNP): real;
(* Programs using routine FNC must define the type
TYPE
   RealArrayNP = ARRAY [1..ndim] OF real;
in the main routine, where ndim is the dimension of vector x. *)
VAR
   f: real;
   i: integer;
BEGIN
   f := 0.0;
   FOR i := 1 TO 3 DO f := f+sqr(x[i]-1.0);
   fnc := f
END;
(*$I F1DIM.PAS *)
FUNCTION fx(x: real): real;
BEGIN
   fx := f1dim(x)
END;
(*$I SCRSHO.PAS *)
BEGIN
   p[1] := 0.0; p[2] := 0.0; p[3] := 0.0;
   LinminNcom := ndim;
   writeln;
   writeln('Enter vector direction along which to');
   writeln('plot the function. Minimum is in the');
   writeln('direction 1.0 1.0 1.0 - enter x y z:');
   read(xi[1],xi[2],xi[3]);
   writeln;
```

```
   FOR j := 1 TO ndim DO BEGIN
      LinminPcom[j] := p[j];
      LinminXicom[j] := xi[j]
   END;
   scrsho
END.
```

frprmn is another multidimensional minimizer that relies on the one-dimensional minimizations of linmin. It works, however, via the Fletcher-Reeves-Polak-Ribiere method and requires that routines be supplied for calculating both the function and its gradient. Sample program d10r9, for example, uses

$$\text{fnc}(x, y, z) = 1.0 - J_0(x - \tfrac{1}{2})J_0(y - \tfrac{1}{2})J_0(z - \tfrac{1}{2})$$

and

$$\frac{\partial\,\text{fnc}}{\partial x} = J_1(x - \tfrac{1}{2})J_0(y - \tfrac{1}{2})J_0(z - \tfrac{1}{2})$$

etc. A number of trial starting vectors are used, and each time, frprmn manages to find the minimum at $(1/2, 1/2, 1/2)$.

```
PROGRAM d10r9(input,output);
(* driver for routine FRPRMN *)
(*$I MODFILE.PAS *)
CONST
   ndim = 3;
   ftol = 1.0e-6;
   pio2 = 1.5707963;
TYPE
   RealArrayNP = ARRAY [1..ndim] OF real;
VAR
   LinminNcom: integer;
   LinminPcom,LinminXicom: RealArrayNP;
   angl,fret: real;
   iter,k: integer;
   p: RealArrayNP;
(*$I BESSJ0.PAS *)
(*$I BESSJ1.PAS *)
FUNCTION fnc(VAR x: RealArrayNP): real;
BEGIN
   fnc := 1.0-bessj0(x[1]-0.5)*bessj0(x[2]-0.5)*bessj0(x[3]-0.5)
END;
PROCEDURE dfnc(VAR x,df: RealArrayNP);
BEGIN
   df[1] := bessj1(x[1]-0.5)*bessj0(x[2]-0.5)*bessj0(x[3]-0.5);
   df[2] := bessj0(x[1]-0.5)*bessj1(x[2]-0.5)*bessj0(x[3]-0.5);
   df[3] := bessj0(x[1]-0.5)*bessj0(x[2]-0.5)*bessj1(x[3]-0.5)
END;
(*$I F1DIM.PAS *)
(*$I MNBRAK.PAS *)
(*$I BRENT.PAS *)
(*$I LINMIN.PAS *)
(*$I FRPRMN.PAS *)
BEGIN
   writeln('Program finds the minimum of a function');
   writeln('with different trial starting vectors.');
   writeln('True minimum is (0.5,0.5,0.5)');
```

```
      FOR k := 0 TO 4 DO BEGIN
         angl := pio2*k/4.0;
         p[1] := 2.0*cos(angl);
         p[2] := 2.0*sin(angl);
         p[3] := 0.0;
         writeln;
         writeln('Starting vector: (',
             p[1]:6:4,',',p[2]:6:4,',',p[3]:6:4,')');
         frprmn(p,ndim,ftol,iter,fret);
         writeln('Iterations:',iter:3);
         writeln('Solution vector: (',
             p[1]:6:4,',',p[2]:6:4,',',p[3]:6:4,')');
         writeln('Func. value at solution',fret:14)
      END
END.
```

Completeness requires that we provide a sample program for **df1dim**, which is presented in *Numerical Recipes* as a routine for converting the N-dimensional gradient procedure to one that provides the first derivative of the function along a specified line in N-dimensional space. It is exactly analogous to **f1dim** and the program **d10r10** is the same.

```
PROGRAM d10r10(input,output);
(* driver for routine DF1DIM *)
(*$I MODFILE.PAS *)
CONST
   ndim = 3;
TYPE
   RealArrayNP = ARRAY [1..ndim] OF real;
VAR
   i,j,LinminNcom: integer;
   p,xi: RealArrayNP;
   LinminPcom,LinminXicom: RealArrayNP;
PROCEDURE dfnc(VAR x,df: RealArrayNP);
VAR
   i: integer;
BEGIN
   FOR i := 1 TO 3 DO df[i] := sqr(x[i]-1.0)
END;
(*$I DF1DIM.PAS *)
FUNCTION fx(x: real): real;
BEGIN
   fx := df1dim(x)
END;
(*$I SCRSHO.PAS *)
BEGIN
   p[1] := 0.0; p[2] := 0.0; p[3] := 0.0;
   LinminNcom := ndim;
   writeln;
   writeln('Enter vector direction along which to');
   writeln('plot the function. Minimum is in the');
   writeln('direction 1.0 1.0 1.0 - enter x y z:');
   read(xi[1],xi[2],xi[3]);
   writeln;
   FOR j := 1 TO ndim DO BEGIN
      LinminPcom[j] := p[j];
      LinminXicom[j] := xi[j]
```

```
      END;
      scrsho
END.
```

dfpmin implements the Broyden-Fletcher-Goldfarb-Shanno variant of the David-on-Fletcher-Powell minimization by variable metric methods. It requires somewhat more intermediate storage than the preceding routine and is not considered superior in other ways. However, it is a popular method. Sample program d10r11 works just as did the program for frprmn, including the fact that it requires a procedure for calculation of the derivative.

```
PROGRAM d10r11(input,output);
(* driver for routine DFPMIN *)
(*$I MODFILE.PAS *)
CONST
   ndim = 3;
   ftol = 1.0e-6;
   pio2 = 1.5707963;
TYPE
   RealArrayNP = ARRAY [1..ndim] OF real;
   RealArrayNPbyNP = ARRAY [1..ndim,1..ndim] OF real;
VAR
   LinminNcom: integer;
   LinminPcom,LinminXicom: RealArrayNP;
   angl,fret: real;
   iter,k: integer;
   p: RealArrayNP;
(*$I BESSJ0.PAS *)
(*$I BESSJ1.PAS *)
FUNCTION fnc(VAR x: RealArrayNP): real;
BEGIN
   fnc := 1.0-bessj0(x[1]-0.5)*bessj0(x[2]-0.5)*bessj0(x[3]-0.5)
END;
PROCEDURE dfnc(VAR x,df: RealArrayNP);
BEGIN
   df[1] := bessj1(x[1]-0.5)*bessj0(x[2]-0.5)*bessj0(x[3]-0.5);
   df[2] := bessj0(x[1]-0.5)*bessj1(x[2]-0.5)*bessj0(x[3]-0.5);
   df[3] := bessj0(x[1]-0.5)*bessj0(x[2]-0.5)*bessj1(x[3]-0.5)
END;
(*$I F1DIM.PAS *)
(*$I MNBRAK.PAS *)
(*$I BRENT.PAS *)
(*$I LINMIN.PAS *)
(*$I DFPMIN.PAS *)
BEGIN
   writeln('Program finds the minimum of a function');
   writeln('with different trial starting vectors.');
   writeln('True minimum is (0.5,0.5,0.5)');
   FOR k := 0 TO 4 DO BEGIN
      angl := pio2*k/4.0;
      p[1] := 2.0*cos(angl);
      p[2] := 2.0*sin(angl);
      p[3] := 0.0;
      writeln;
      writeln('Starting vector: (',
         p[1]:6:4,',',p[2]:6:4,',',p[3]:6:4,')');
```

```
        dfpmin(p,ndim,ftol,iter,fret);
        writeln('Iterations:',iter:3);
        writeln('Solution vector: (',
            p[1]:6:4,',',p[2]:6:4,',',p[3]:6:4,')');
        writeln('Func. value at solution',fret:14)
    END
END.
```

simplx is a procedure for dealing with problems in linear programming. In these problems the goal is to maximize a linear combination of N variables, subject to the constraint that none be negative, and that as a group they satisfy a number of other constraints. In order to clarify the subject, *Numerical Recipes* presents a sample problem in equations (10.8.6) and (10.8.7), translating the problem into tableau format in (10.8.18), and presenting a solution in equation (10.8.19). Sample program d10r12 carries out the analysis that leads to this solution.

```
PROGRAM d10r12(input,output);
(* driver for routine SIMPLX *)
(* incorporates examples discussed in text *)
(*$I MODFILE.PAS *)
CONST
   n = 4;
   m = 4;
   np = 5;        (* np >= n+1 *)
   mp = 6;        (* mp >= m+2 *)
   m1 = 2;        (* m1+m2+m3=m *)
   m2 = 1;
   m3 = 1;
   nm1m2 = 7;     (* nm1m2=n+m1+m2 *)
TYPE
   CharArray2 = PACKED ARRAY [1..2] OF char;
   RealArrayMPbyNP = ARRAY [1..mp,1..np] OF real;
   IntegerArrayN = ARRAY [1..n] OF integer;
   IntegerArrayM = ARRAY [1..m] OF integer;
   IntegerArrayNP = ARRAY [1..np] OF integer;
VAR
   i,icase,j: integer;
   rite: boolean;
   izrov: IntegerArrayN;
   iposv: IntegerArrayM;
   a: RealArrayMPbyNP;
   txt: ARRAY [1..nm1m2] OF CharArray2;
(*$I SIMPLX.PAS *)
BEGIN
   txt[1] := 'x1'; txt[2] := 'x2'; txt[3] := 'x3';
   txt[4] := 'x4'; txt[5] := 'y1'; txt[6] := 'y2';
   txt[7] := 'y3';
   a[1,1] := 0.0; a[1,2] := 1.0; a[1,3] := 1.0;
   a[1,4] := 3.0; a[1,5] := -0.5;
   a[2,1] := 740.0; a[2,2] := -1.0; a[2,3] := 0.0;
   a[2,4] := -2.0; a[2,5] := 0.0;
   a[3,1] := 0.0; a[3,2] := 0.0; a[3,3] := -2.0;
   a[3,4] := 0.0; a[3,5] := 7.0;
   a[4,1] := 0.5; a[4,2] := 0.0; a[4,3] := -1.0;
   a[4,4] := 1.0; a[4,5] := -2.0;
   a[5,1] := 9.0; a[5,2] := -1.0; a[5,3] := -1.0;
```

```
      a[5,4] := -1.0; a[5,5] := -1.0;
      simplx(a,m,n,m1,m2,m3,icase,izrov,iposv);
      writeln;
      IF icase = 1 THEN
         writeln('unbounded objective function')
      ELSE IF icase = -1 THEN
         writeln('no solutions satisfy constraints given')
      ELSE BEGIN
         write(' ':11);
         FOR i := 1 TO n DO
            IF izrov[i] <= nm1m2 THEN write(txt[izrov[i]]:10);
         writeln;
         FOR i := 1 TO m+1 DO BEGIN
            IF i = 1 THEN BEGIN
               write(' ');
               rite := true
            END ELSE IF iposv[i-1] <= nm1m2 THEN BEGIN
               write(txt[iposv[i-1]]);
               rite := true
            END ELSE
               rite := false;
            IF rite THEN BEGIN
               write(a[i,1]:10:2);
               FOR j := 2 TO n+1 DO
                  IF izrov[j-1] <= nm1m2 THEN write(a[i,j]:10:2);
               writeln
            END
         END
      END
END.
```

anneal is a procedure for solving the travelling salesman problem—a problem that is included as a demonstration of the use of simulated annealing. Sample program d10r13 has the function of setting up the initial route for the salesman and printing final results. For each of ncity=10 cities, it chooses random coordinates x[i],y[i] using routine ran3, and puts an entry for each city in the array iptr[i]. The array indicates the order in which the cities will be visited. On the originally specified path, the cities are in the order i=1,...,10 so the sample program initially takes iptr[i]=i. (It is assumed that the salesman will return to the first city after visiting the last.) A call is then made to anneal, which attempts to find the shortest alternative route and record it in the array. After finding a path that resists further improvement, the driver lists the modified itinerary.

```
PROGRAM D10R13(input,output);
(*$I MODFILE.PAS *)
CONST
   ncity = 10;
TYPE
   RealArray55 = ARRAY [1..55] OF real;
   RealArrayNCITY = ARRAY [1..ncity] OF real;
   IntegerArrayNCITY = ARRAY [1..ncity] OF integer;
VAR
   Ran3Inext,Ran3Inextp,MetropJdum,idum,i,ii: integer;
   Ran3Ma: RealArray55;
   x,y: RealArrayNCITY;
   iorder: IntegerArrayNCITY;
```

```
(*$I RAN3.PAS *)
(*$I IRBIT1.PAS *)
(*$I ANNEAL.PAS *)
BEGIN
   MetropJdum := 1;
   idum := -1;
   FOR i := 1 TO ncity DO BEGIN
      x[i] := ran3(idum);
      y[i] := ran3(idum);
      iorder[i] := i
   END;
   anneal(x,y,iorder,ncity);
   writeln('*** System Frozen ***');
   writeln('Final path:');
   writeln(' ':3,'city',' ':6,'x',' ':9,'y');
   FOR i := 1 TO ncity DO BEGIN
      ii := iorder[i];
      writeln(ii:4,x[ii]:10:4,y[ii]:10:4)
   END
END.
```

Chapter 11: Eigensystems

In Chapter 11 of *Numerical Recipes*, we deal with the problem of finding eigenvectors and eigenvalues of matrices, first dealing with symmetric matrices, and then with more general cases. For real symmetric matrices of small-to-moderate size, the routine jacobi is recommended as a simple and foolproof scheme of finding eigenvalues and eigenvectors. Routine eigsrt may be used to reorder the output of jacobi into descending order of eigenvalue. A more efficient (but operationally more complicated) procedure is to reduce the symmetric matrix to tridiagonal form before doing the eigenvalue analysis. tred2 uses the Householder scheme to perform this reduction and is used in conjunction with tqli. tqli determines the eigenvalues and eigenvectors of a real, symmetric, tridiagonal matrix.

For nonsymmetric matrices, we offer only routines for finding eigenvalues, and not eigenvectors. To ameliorate problems with roundoff error, balanc makes the corresponding rows and columns of the matrix have comparable norms while leaving eigenvalues unchanged. Then the matrix is reduced to Hessenberg form by Gaussian elimination using elmhes. Finally hqr applies the QR algorithm to find the eigenvalues of the Hessenberg matrix.

$$\star \quad \star \quad \star \quad \star$$

jacobi is a reliable scheme for finding both the eigenvalues and eigenvectors of a symmetric matrix. It is not the most efficient scheme available, but it is simple and trustworthy, and it is recommended for problems of small-to-moderate order. Sample program d11r1 defines three matrices a,b,c for use by jacobi. They are of order 3, 5, and 10 respectively. To check the index handling in jacobi, they are, each in turn, sent to jacobi with np=10 (the physical size of the matrix) and n=3, 5 or 10 (the logical array dimension). For each matrix, the eigenvalues and eigenvectors are reported. Then, an eigenvector test takes place in which the original matrix is applied to the purported eigenvector, and the ratio of the result to the vector itself is found. The ratio should, of course, be the eigenvalue.

```
PROGRAM d11r1(input,output);
(* driver for routine JACOBI *)
(*$I MODFILE.PAS *)
CONST
   np = 10;
   nmat = 3;
TYPE
   IntegerArray3 = ARRAY [1..3] OF integer;
   RealArrayNPbyNP = ARRAY [1..np,1..np] OF real;
   RealArrayNP = ARRAY [1..np] OF real;
```

```
VAR
    i,j,k,kk,l,ll,nrot: integer;
    a,b,c,v: RealArrayNPbyNP;
    d,r: RealArrayNP;
    num: IntegerArray3;
(*$I JACOBI.PAS *)
BEGIN
    num[1] := 3; num[2] := 5; num[3] := 10;
    a[1,1] := 1.0; a[1,2] := 2.0; a[1,3] := 3.0;
    a[2,1] := 2.0; a[2,2] := 2.0; a[2,3] := 3.0;
    a[3,1] := 3.0; a[3,2] := 3.0; a[3,3] := 3.0;
    b[1,1] := -2.0; b[1,2] := -1.0; b[1,3] := 0.0;
    b[1,4] := 1.0; b[1,5] := 2.0;
    b[2,1] := -1.0; b[2,2] := -1.0; b[2,3] := 0.0;
    b[2,4] := 1.0; b[2,5] := 2.0;
    b[3,1] := 0.0; b[3,2] := 0.0; b[3,3] := 0.0;
    b[3,4] := 1.0; b[3,5] := 2.0;
    b[4,1] := 1.0; b[4,2] := 1.0; b[4,3] := 1.0;
    b[4,4] := 1.0; b[4,5] := 2.0;
    b[5,1] := 2.0; b[5,2] := 2.0; b[5,3] := 2.0;
    b[5,4] := 2.0; b[5,5] := 2.0;
    c[1,1] := 5.0; c[1,2] := 4.0; c[1,3] := 3.0;
    c[1,4] := 2.0; c[1,5] := 1.0; c[1,6] := 0.0;
    c[1,7] := -1.0; c[1,8] := -2.0; c[1,9] := -3.0;
    c[1,10] := -4.0;
    c[2,1] := 4.0; c[2,2] := 5.0; c[2,3] := 4.0;
    c[2,4] := 3.0; c[2,5] := 2.0; c[2,6] := 1.0;
    c[2,7] := 0.0; c[2,8] := -1.0; c[2,9] := -2.0;
    c[2,10] := -3.0;
    c[3,1] := 3.0; c[3,2] := 4.0; c[3,3] := 5.0;
    c[3,4] := 4.0; c[3,5] := 3.0; c[3,6] := 2.0;
    c[3,7] := 1.0; c[3,8] := 0.0; c[3,9] := -1.0;
    c[3,10] := -2.0;
    c[4,1] := 2.0; c[4,2] := 3.0; c[4,3] := 4.0;
    c[4,4] := 5.0; c[4,5] := 4.0; c[4,6] := 3.0;
    c[4,7] := 2.0; c[4,8] := 1.0; c[4,9] := 0.0;
    c[4,10] := -1.0;
    c[5,1] := 1.0; c[5,2] := 2.0; c[5,3] := 3.0;
    c[5,4] := 4.0; c[5,5] := 5.0; c[5,6] := 4.0;
    c[5,7] := 3.0; c[5,8] := 2.0; c[5,9] := 1.0;
    c[5,10] := 0.0;
    c[6,1] := 0.0; c[6,2] := 1.0; c[6,3] := 2.0;
    c[6,4] := 3.0; c[6,5] := 4.0; c[6,6] := 5.0;
    c[6,7] := 4.0; c[6,8] := 3.0; c[6,9] := 2.0;
    c[6,10] := 1.0;
    c[7,1] := -1.0; c[7,2] := 0.0; c[7,3] := 1.0;
    c[7,4] := 2.0; c[7,5] := 3.0; c[7,6] := 4.0;
    c[7,7] := 5.0; c[7,8] := 4.0; c[7,9] := 3.0;
    c[7,10] := 2.0;
    c[8,1] := -2.0; c[8,2] := -1.0; c[8,3] := 0.0;
    c[8,4] := 1.0; c[8,5] := 2.0; c[8,6] := 3.0;
    c[8,7] := 4.0; c[8,8] := 5.0; c[8,9] := 4.0;
    c[8,10] := 3.0;
    c[9,1] := -3.0; c[9,2] := -2.0; c[9,3] := -1.0;
    c[9,4] := 0.0; c[9,5] := 1.0; c[9,6] := 2.0;
    c[9,7] := 3.0; c[9,8] := 4.0; c[9,9] := 5.0;
    c[9,10] := 4.0;
```

```
    c[10,1] := -4.0; c[10,2] := -3.0; c[10,3] := -2.0;
    c[10,4] := -1.0; c[10,5] := 0.0; c[10,6] := 1.0;
    c[10,7] := 2.0; c[10,8] := 3.0; c[10,9] := 4.0;
    c[10,10] := 5.0;
    FOR i := 1 TO nmat DO BEGIN
        IF i = 1 THEN
            jacobi(a,num[i],d,v,nrot)
        ELSE IF i = 2 THEN
            jacobi(b,num[i],d,v,nrot)
        ELSE IF i = 3 THEN
            jacobi(c,num[i],d,v,nrot);
        writeln('matrix number',i:2);
        writeln('number of jacobi rotations:',nrot:3);
        writeln('eigenvalues:');
        FOR j := 1 TO num[i] DO BEGIN
            write(d[j]:12:6);
            IF j MOD 5 = 0 THEN writeln
        END;
        writeln;
        writeln('eigenvectors:');
        FOR j := 1 TO num[i] DO BEGIN
            writeln('number':9,j:3);
            FOR k := 1 TO num[i] DO BEGIN
                write(v[k,j]:12:6);
                IF k MOD 5 = 0 THEN writeln
            END;
            writeln
        END;
(* eigenvector test *)
        writeln('eigenvector test');
        FOR j := 1 TO num[i] DO BEGIN
            FOR l := 1 TO num[i] DO BEGIN
                r[l] := 0.0;
                FOR k := 1 TO num[i] DO BEGIN
                    IF k > l THEN BEGIN
                        kk := l;
                        ll := k;
                    END ELSE BEGIN
                        kk := k;
                        ll := l;
                    END;
                    IF i = 1 THEN
                        r[l] := r[l]+a[ll,kk]*v[k,j]
                    ELSE IF i = 2 THEN
                        r[l] := r[l]+b[ll,kk]*v[k,j]
                    ELSE IF i = 3 THEN
                        r[l] := r[l]+c[ll,kk]*v[k,j]
                END
            END;
            writeln('vector number',j:3);
            writeln('vector':11,'mtrx*vec.':14,'ratio':10);
            FOR l := 1 TO num[i] DO
                writeln(v[l,j]:12:6,r[l]:12:6,r[l]/v[l,j]:12:6)
        END;
        writeln('press return to continue...');
        readln
    END
```

END.

eigsrt reorders the output of jacobi so that the eigenvectors are in the order of decreasing eigenvalue. Sample program d11r2 uses matrix c from the previous program to illustrate. This 10×10 matrix is passed to jacobi and the ten eigenvectors are found. They are printed, along with their eigenvalues, in the order that jacobi returns them. Then the matrices d and v from jacobi, which contain the eigenvalues and eigenvectors, are passed to eigsrt, and ought to return in descending order of eigenvalue. The result is printed for inspection.

```
PROGRAM d11r2(input,output);
(* driver for routine EIGSRT *)
(*$I MODFILE.PAS *)
CONST
   np = 10;
TYPE
   RealArrayNPbyNP = ARRAY [1..np,1..np] OF real;
   RealArrayNP = ARRAY [1..np] OF real;
VAR
   i,j,nrot: integer;
   c,v: RealArrayNPbyNP;
   d: RealArrayNP;
(*$I JACOBI.PAS *)
(*$I EIGSRT.PAS *)
BEGIN
   c[1,1] := 5.0; c[1,2] := 4.0; c[1,3] := 3.0;
   c[1,4] := 2.0; c[1,5] := 1.0; c[1,6] := 0.0;
   c[1,7] := -1.0; c[1,8] := -2.0; c[1,9] := -3.0;
   c[1,10] := -4.0;
   c[2,1] := 4.0; c[2,2] := 5.0; c[2,3] := 4.0;
   c[2,4] := 3.0; c[2,5] := 2.0; c[2,6] := 1.0;
   c[2,7] := 0.0; c[2,8] := -1.0; c[2,9] := -2.0;
   c[2,10] := -3.0;
   c[3,1] := 3.0; c[3,2] := 4.0; c[3,3] := 5.0;
   c[3,4] := 4.0; c[3,5] := 3.0; c[3,6] := 2.0;
   c[3,7] := 1.0; c[3,8] := 0.0; c[3,9] := -1.0;
   c[3,10] := -2.0;
   c[4,1] := 2.0; c[4,2] := 3.0; c[4,3] := 4.0;
   c[4,4] := 5.0; c[4,5] := 4.0; c[4,6] := 3.0;
   c[4,7] := 2.0; c[4,8] := 1.0; c[4,9] := 0.0;
   c[4,10] := -1.0;
   c[5,1] := 1.0; c[5,2] := 2.0; c[5,3] := 3.0;
   c[5,4] := 4.0; c[5,5] := 5.0; c[5,6] := 4.0;
   c[5,7] := 3.0; c[5,8] := 2.0; c[5,9] := 1.0;
   c[5,10] := 0.0;
   c[6,1] := 0.0; c[6,2] := 1.0; c[6,3] := 2.0;
   c[6,4] := 3.0; c[6,5] := 4.0; c[6,6] := 5.0;
   c[6,7] := 4.0; c[6,8] := 3.0; c[6,9] := 2.0;
   c[6,10] := 1.0;
   c[7,1] := -1.0; c[7,2] := 0.0; c[7,3] := 1.0;
   c[7,4] := 2.0; c[7,5] := 3.0; c[7,6] := 4.0;
   c[7,7] := 5.0; c[7,8] := 4.0; c[7,9] := 3.0;
   c[7,10] := 2.0;
   c[8,1] := -2.0; c[8,2] := -1.0; c[8,3] := 0.0;
   c[8,4] := 1.0; c[8,5] := 2.0; c[8,6] := 3.0;
   c[8,7] := 4.0; c[8,8] := 5.0; c[8,9] := 4.0;
```

```
c[8,10] := 3.0;
c[9,1] := -3.0; c[9,2] := -2.0; c[9,3] := -1.0;
c[9,4] := 0.0; c[9,5] := 1.0; c[9,6] := 2.0;
c[9,7] := 3.0; c[9,8] := 4.0; c[9,9] := 5.0;
c[9,10] := 4.0;
c[10,1] := -4.0; c[10,2] := -3.0; c[10,3] := -2.0;
c[10,4] := -1.0; c[10,5] := 0.0; c[10,6] := 1.0;
c[10,7] := 2.0; c[10,8] := 3.0; c[10,9] := 4.0;
c[10,10] := 5.0;
jacobi(c,np,d,v,nrot);
writeln('unsorted eigenvectors:');
FOR i := 1 TO np DO BEGIN
   writeln('eigenvalue',i:3,'   := ',d[i]:12:6);
   writeln('eigenvector:');
   FOR j := 1 TO np DO BEGIN
      write(v[j,i]:12:6);
      IF j MOD 5 = 0 THEN writeln
   END;
   writeln
END;
writeln;
writeln('****** sorting ******');
writeln;
eigsrt(d,v,np);
writeln('sorted eigenvectors:');
FOR i := 1 TO np DO BEGIN
   writeln('eigenvalue',i:3,'   := ',d[i]:12:6);
   writeln('eigenvector:');
   FOR j := 1 TO np DO BEGIN
      write(v[j,i]:12:6);
      IF j MOD 5 = 0 THEN writeln
   END;
   writeln
END
END.
```

tred2 reduces a real symmetric matrix to tridiagonal form. Sample program tred2 again uses matrix c from the earlier programs, and copies it into matrix a. Matrix a is sent to tred2, while c is saved for a check of the transformation matrix that tred2 returns in a. The program prints the diagonal and off-diagonal elements of the reduced matrix. It then forms the matrix f defined by $F = A^T C A$ to prove that f is tridiagonal and that the listed diagonal and off-diagonal elements are correct.

```
PROGRAM d11r3(input,output);
(* driver for routine TRED2 *)
(*$I MODFILE.PAS *)
CONST
   np = 10;
TYPE
   RealArrayNP = ARRAY [1..np] OF real;
   RealArrayNPbyNP = ARRAY [1..np,1..np] OF real;
VAR
   i,j,k,l,m: integer;
   a,c,f: RealArrayNPbyNP;
   d,e: RealArrayNP;
(*$I TRED2.PAS *)
```

```
BEGIN
    c[1,1] := 5.0; c[1,2] := 4.0; c[1,3] := 3.0;
    c[1,4] := 2.0; c[1,5] := 1.0; c[1,6] := 0.0;
    c[1,7] := -1.0; c[1,8] := -2.0; c[1,9] := -3.0;
    c[1,10] := -4.0;
    c[2,1] := 4.0; c[2,2] := 5.0; c[2,3] := 4.0;
    c[2,4] := 3.0; c[2,5] := 2.0; c[2,6] := 1.0;
    c[2,7] := 0.0; c[2,8] := -1.0; c[2,9] := -2.0;
    c[2,10] := -3.0;
    c[3,1] := 3.0; c[3,2] := 4.0; c[3,3] := 5.0;
    c[3,4] := 4.0; c[3,5] := 3.0; c[3,6] := 2.0;
    c[3,7] := 1.0; c[3,8] := 0.0; c[3,9] := -1.0;
    c[3,10] := -2.0;
    c[4,1] := 2.0; c[4,2] := 3.0; c[4,3] := 4.0;
    c[4,4] := 5.0; c[4,5] := 4.0; c[4,6] := 3.0;
    c[4,7] := 2.0; c[4,8] := 1.0; c[4,9] := 0.0;
    c[4,10] := -1.0;
    c[5,1] := 1.0; c[5,2] := 2.0; c[5,3] := 3.0;
    c[5,4] := 4.0; c[5,5] := 5.0; c[5,6] := 4.0;
    c[5,7] := 3.0; c[5,8] := 2.0; c[5,9] := 1.0;
    c[5,10] := 0.0;
    c[6,1] := 0.0; c[6,2] := 1.0; c[6,3] := 2.0;
    c[6,4] := 3.0; c[6,5] := 4.0; c[6,6] := 5.0;
    c[6,7] := 4.0; c[6,8] := 3.0; c[6,9] := 2.0;
    c[6,10] := 1.0;
    c[7,1] := -1.0; c[7,2] := 0.0; c[7,3] := 1.0;
    c[7,4] := 2.0; c[7,5] := 3.0; c[7,6] := 4.0;
    c[7,7] := 5.0; c[7,8] := 4.0; c[7,9] := 3.0;
    c[7,10] := 2.0;
    c[8,1] := -2.0; c[8,2] := -1.0; c[8,3] := 0.0;
    c[8,4] := 1.0; c[8,5] := 2.0; c[8,6] := 3.0;
    c[8,7] := 4.0; c[8,8] := 5.0; c[8,9] := 4.0;
    c[8,10] := 3.0;
    c[9,1] := -3.0; c[9,2] := -2.0; c[9,3] := -1.0;
    c[9,4] := 0.0; c[9,5] := 1.0; c[9,6] := 2.0;
    c[9,7] := 3.0; c[9,8] := 4.0; c[9,9] := 5.0;
    c[9,10] := 4.0;
    c[10,1] := -4.0; c[10,2] := -3.0; c[10,3] := -2.0;
    c[10,4] := -1.0; c[10,5] := 0.0; c[10,6] := 1.0;
    c[10,7] := 2.0; c[10,8] := 3.0; c[10,9] := 4.0;
    c[10,10] := 5.0;
    FOR i := 1 TO np DO
        FOR j := 1 TO np DO a[i,j] := c[i,j];
    tred2(a,np,d,e);
    writeln('diagonal elements');
    FOR i := 1 TO np DO BEGIN
        write(d[i]:12:6);
        IF i MOD 5 = 0 THEN writeln
    END;
    writeln('off-diagonal elements');
    FOR i := 2 TO np DO BEGIN
        write(e[i]:12:6);
        IF i MOD 5 = 0 THEN writeln
    END;
(* check transformation matrix *)
    FOR j := 1 TO np DO BEGIN
        FOR k := 1 TO np DO BEGIN
```

```
            f[j,k] := 0.0;
            FOR l := 1 TO np DO
                FOR m := 1 TO np DO
                    f[j,k] := f[j,k]+a[l,j]*c[l,m]*a[m,k]
        END
    END;
(* how does it look? *)
    writeln('tridiagonal matrix');
    FOR i := 1 TO np DO BEGIN
        FOR j := 1 TO np DO write(f[i,j]:7:2);
        writeln
    END
END.
```

tqli finds the eigenvectors and eigenvalues for a real, symmetric, tridiagonal matrix. Sample program d11r4 operates with matrix c again, and uses tred2 to reduce it to tridiagonal form as before. More specifically, c is copied into matrix a, which is sent to tred2. From tred2 come two vectors d,e which are the diagonal and subdiagonal elements of the tridiagonal matrix. d and e are made arguments of tqli, as is a, the returned transformation matrix from tred2. On output from tqli, d is replaced with eigenvalues, and a with corresponding eigenvectors. These are checked as in the program for jacobi. That is, the original matrix c is applied to each eigenvector, and the result is divided (element by element) by the eigenvector. Look for a result equal to the eigenvalue. (Note: in some cases, the vector element is zero or nearly so. These cases are flagged with the words "div. by zero".)

```
PROGRAM d11r4(input,output);
(* driver for routine TQLI *)
(*$I MODFILE.PAS *)
CONST
    np = 10;
    tiny = 1.0e-6;
TYPE
    RealArrayNP = ARRAY [1..np] OF real;
    RealArrayNPbyNP = ARRAY [1..np,1..np] OF real;
VAR
    i,j,k: integer;
    a,c: RealArrayNPbyNP;
    d,e,f: RealArrayNP;
(*$I TRED2.PAS *)
(*$I TQLI.PAS *)
BEGIN
    c[1,1] := 5.0; c[1,2] := 4.0; c[1,3] := 3.0;
    c[1,4] := 2.0; c[1,5] := 1.0; c[1,6] := 0.0;
    c[1,7] := -1.0; c[1,8] := -2.0; c[1,9] := -3.0;
    c[1,10] := -4.0;
    c[2,1] := 4.0; c[2,2] := 5.0; c[2,3] := 4.0;
    c[2,4] := 3.0; c[2,5] := 2.0; c[2,6] := 1.0;
    c[2,7] := 0.0; c[2,8] := -1.0; c[2,9] := -2.0;
    c[2,10] := -3.0;
    c[3,1] := 3.0; c[3,2] := 4.0; c[3,3] := 5.0;
    c[3,4] := 4.0; c[3,5] := 3.0; c[3,6] := 2.0;
    c[3,7] := 1.0; c[3,8] := 0.0; c[3,9] := -1.0;
    c[3,10] := -2.0;
    c[4,1] := 2.0; c[4,2] := 3.0; c[4,3] := 4.0;
    c[4,4] := 5.0; c[4,5] := 4.0; c[4,6] := 3.0;
```

```
    c[4,7]  := 2.0; c[4,8]  := 1.0; c[4,9]  := 0.0;
    c[4,10] := -1.0;
    c[5,1]  := 1.0; c[5,2]  := 2.0; c[5,3]  := 3.0;
    c[5,4]  := 4.0; c[5,5]  := 5.0; c[5,6]  := 4.0;
    c[5,7]  := 3.0; c[5,8]  := 2.0; c[5,9]  := 1.0;
    c[5,10] := 0.0;
    c[6,1]  := 0.0; c[6,2]  := 1.0; c[6,3]  := 2.0;
    c[6,4]  := 3.0; c[6,5]  := 4.0; c[6,6]  := 5.0;
    c[6,7]  := 4.0; c[6,8]  := 3.0; c[6,9]  := 2.0;
    c[6,10] := 1.0;
    c[7,1]  := -1.0; c[7,2]  := 0.0; c[7,3]  := 1.0;
    c[7,4]  := 2.0; c[7,5]  := 3.0; c[7,6]  := 4.0;
    c[7,7]  := 5.0; c[7,8]  := 4.0; c[7,9]  := 3.0;
    c[7,10] := 2.0;
    c[8,1]  := -2.0; c[8,2]  := -1.0; c[8,3]  := 0.0;
    c[8,4]  := 1.0; c[8,5]  := 2.0; c[8,6]  := 3.0;
    c[8,7]  := 4.0; c[8,8]  := 5.0; c[8,9]  := 4.0;
    c[8,10] := 3.0;
    c[9,1]  := -3.0; c[9,2]  := -2.0; c[9,3]  := -1.0;
    c[9,4]  := 0.0; c[9,5]  := 1.0; c[9,6]  := 2.0;
    c[9,7]  := 3.0; c[9,8]  := 4.0; c[9,9]  := 5.0;
    c[9,10] := 4.0;
    c[10,1]  := -4.0; c[10,2]  := -3.0; c[10,3]  := -2.0;
    c[10,4]  := -1.0; c[10,5]  := 0.0; c[10,6]  := 1.0;
    c[10,7]  := 2.0; c[10,8]  := 3.0; c[10,9]  := 4.0;
    c[10,10] := 5.0;
    FOR i := 1 TO np DO
       FOR j := 1 TO np DO a[i,j] := c[i,j];
    tred2(a,np,d,e);
    tqli(d,e,np,a);
    writeln('eigenvectors for a real symmetric matrix');
    FOR i := 1 TO np DO BEGIN
       FOR j := 1 TO np DO BEGIN
          f[j] := 0.0;
          FOR k := 1 TO np DO f[j] := f[j]+c[j,k]*a[k,i]
       END;
       writeln('eigenvalue',i:3,'   := ',d[i]:10:6);
       writeln('vector':11,'mtrx*vect.':14,'ratio':9);
       FOR j := 1 TO np DO
          IF abs(a[j,i]) < tiny THEN
             writeln(a[j,i]:12:6,f[j]:12:6,'div. by 0':12)
          ELSE
             writeln(a[j,i]:12:6,f[j]:12:6,f[j]/a[j,i]:12:6);
       writeln('press enter to continue...');
       readln
    END
END.
```

balanc reduces error in eigenvalue problems involving non-symmetric matrices. It does this by adjusting corresponding rows and columns to have comparable norms, without changing eigenvalues. Sample program d11r5 prepares the following array a for balanc

$$\begin{pmatrix} 1 & 100 & 1 & 100 & 1 \\ 1 & 1 & 1 & 1 & 1 \\ 1 & 100 & 1 & 100 & 1 \\ 1 & 1 & 1 & 1 & 1 \\ 1 & 100 & 1 & 100 & 1 \end{pmatrix}$$

The norms of the five rows and five columns are printed out. It is clear from the array that three of the rows and two of the columns have much larger norms than the others. After balancing with balanc, the norms are recalculated, and this time the row, column pairs should be much more nearly equal.

```
PROGRAM d11r5(input,output);
(* driver for routine BALANC *)
(*$I MODFILE.PAS *)
CONST
   np = 5;
TYPE
   RealArrayNP = ARRAY [1..np] OF real;
   RealArrayNPbyNP = ARRAY [1..np,1..np] OF real;
VAR
   i,j: integer;
   a: RealArrayNPbyNP;
   c,r: RealArrayNP;
(*$I BALANC.PAS *)
BEGIN
   a[1,1] := 1.0; a[1,2] := 100.0; a[1,3] := 1.0;
   a[1,4] := 100.0; a[1,5] := 1.0;
   a[2,1] := 1.0; a[2,2] := 1.0; a[2,3] := 1.0;
   a[2,4] := 1.0; a[2,5] := 1.0;
   a[3,1] := 1.0; a[3,2] := 100.0; a[3,3] := 1.0;
   a[3,4] := 100.0; a[3,5] := 1.0;
   a[4,1] := 1.0; a[4,2] := 1.0; a[4,3] := 1.0;
   a[4,4] := 1.0; a[4,5] := 1.0;
   a[5,1] := 1.0; a[5,2] := 100.0; a[5,3] := 1.0;
   a[5,4] := 100.0; a[5,5] := 1.0;
(* write norms *)
   FOR i := 1 TO np DO BEGIN
      r[i] := 0.0;
      c[i] := 0.0;
      FOR j := 1 TO np DO BEGIN
         r[i] := r[i]+abs(a[i,j]);
         c[i] := c[i]+abs(a[j,i])
      END
   END;
   writeln('rows:');
   FOR i := 1 TO np DO write(r[i]:12:2);
   writeln;
   writeln('columns:');
   FOR i := 1 TO np DO write(c[i]:12:2);
   writeln;
   writeln;
   writeln('***** balancing matrix *****');
   writeln;
   balanc(a,np);
(* write norms *)
   FOR i := 1 TO np DO BEGIN
      r[i] := 0.0;
```

```
        c[i] := 0.0;
        FOR j := 1 TO np DO BEGIN
            r[i] := r[i]+abs(a[i,j]);
            c[i] := c[i]+abs(a[j,i])
        END
    END;
    writeln('rows:');
    FOR i := 1 TO np DO write(r[i]:12:2);
    writeln;
    writeln('columns:');
    FOR i := 1 TO np DO write(c[i]:12:2);
    writeln
END.
```

elmhes reduces a general matrix to Hessenberg form using Gaussian elimination. It is particularly valuable for real, non-symmetric matrices. Sample program d11r6 employs balanc and elmhes to get a non-symmetric and grossly unbalanced matrix into Hessenberg form. The matrix a is

$$\begin{pmatrix} 1 & 2 & 300 & 4 & 5 \\ 2 & 3 & 400 & 5 & 6 \\ 3 & 4 & 5 & 6 & 7 \\ 4 & 5 & 600 & 7 & 8 \\ 5 & 6 & 700 & 8 & 9 \end{pmatrix}$$

After printing the original matrix, the program feeds it to balanc and prints the balanced version. This is submitted to elmhes and the result is printed. Notice that the elements of a with $i > j+1$ are all set to zero by the program, because elmhes returns random values in this part of the matrix. Therefore, you should not attach any importance to the fact that the printed output of the program has Hessenberg form. More important are the contents of the non-zero entries. We include here the expected results for comparison.

Balanced Matrix:

1.00	2.00	37.50	4.00	5.00
2.00	3.00	50.00	5.00	6.00
24.00	32.00	5.00	48.00	56.00
4.00	5.00	75.00	7.00	8.00
5.00	6.00	87.50	8.00	9.00

Reduced to Hessenberg Form:

.1000E+01	.3938E+02	.9618E+01	.3333E+01	.4000E+01
.2400E+02	.2733E+02	.1161E+03	.4800E+02	.4800E+02
.0000E+00	.8551E+02	−.4780E+01	−.1333E+01	−.2000E+01
.0000E+00	.0000E+00	.5188E+01	.1447E+01	.2171E+01
.0000E+00	.0000E+00	.0000E+00	−.9155E−07	.7874E−07

```
PROGRAM d11r6(input,output);
(* driver for routine ELMHES *)
(*$I MODFILE.PAS *)
CONST
    np = 5;
TYPE
    RealArrayNP = ARRAY [1..np] OF real;
    RealArrayNPbyNP = ARRAY [1..np,1..np] OF real;
```

```
VAR
   i,j: integer;
   a: RealArrayNPbyNP;
   c,r: RealArrayNP;
(*$I BALANC.PAS *)
(*$I ELMHES.PAS *)
BEGIN
   a[1,1] := 1.0; a[1,2] := 2.0; a[1,3] := 300.0;
   a[1,4] := 4.0; a[1,5] := 5.0;
   a[2,1] := 2.0; a[2,2] := 3.0; a[2,3] := 400.0;
   a[2,4] := 5.0; a[2,5] := 6.0;
   a[3,1] := 3.0; a[3,2] := 4.0; a[3,3] := 5.0;
   a[3,4] := 6.0; a[3,5] := 7.0;
   a[4,1] := 4.0; a[4,2] := 5.0; a[4,3] := 600.0;
   a[4,4] := 7.0; a[4,5] := 8.0;
   a[5,1] := 5.0; a[5,2] := 6.0; a[5,3] := 700.0;
   a[5,4] := 8.0; a[5,5] := 9.0;
   writeln('***** original matrix *****');
   FOR i := 1 TO np DO BEGIN
      FOR j := 1 TO np DO write(a[i,j]:12:2);
      writeln
   END;
   writeln('***** balance matrix *****');
   balanc(a,np);
   FOR i := 1 TO np DO BEGIN
      FOR j := 1 TO np DO write(a[i,j]:12:2);
      writeln
   END;
   writeln('***** reduce to hessenberg form *****');
   elmhes(a,np);
   FOR j := 1 TO np-2 DO
      FOR i := j+2 TO np DO a[i,j] := 0.0;
   FOR i := 1 TO np DO BEGIN
      FOR j := 1 TO np DO write(' ',a[i,j]:11);
      writeln
   END
END.
```

hqr, finally, is a routine for finding the eigenvalues of a Hessenberg matrix using the QR algorithm. The 5×5 matrix specified in the array a is treated just as you would expect to treat any general real non-symmetric matrix. It is fed to balanc for balancing, to elmhes for reduction to Hessenberg form, and to hqr for eigenvalue determination. The eigenvalues may be complex-valued, and both real and imaginary parts are given. The original matrix has enough strategically placed zeros in it that you should have no trouble finding the eigenvalues by hand. Alternatively, you may check them against the list below:

Matrix:

1.00	2.00	.00	.00	.00
−2.00	3.00	.00	.00	.00
3.00	4.00	50.00	.00	.00
−4.00	5.00	−60.00	7.00	.00
−5.00	6.00	−70.00	8.00	−9.00

Eigenvalues:

#	Real	Imag.
1	.500000E+02	.000000E+00
2	.200000E+01	-.173205E+01
3	.200000E+01	.173205E+01
4	.700000E+01	.000000E+00
5	-.900000E+01	.000000E+00

```pascal
PROGRAM d11r7(input,output);
(* driver for routine HQR *)
(*$I MODFILE.PAS *)
CONST
   np = 5;
TYPE
   RealArrayNPbyNP = ARRAY [1..np,1..np] OF real;
   RealArrayNP = ARRAY [1..np] OF real;
VAR
   i,j: integer;
   a: RealArrayNPbyNP;
   wi,wr: RealArrayNP;
(*$I BALANC.PAS *)
(*$I ELMHES.PAS *)
(*$I HQR.PAS *)
BEGIN
   a[1,1] := 1.0; a[1,2] := 2.0; a[1,3] := 0.0;
   a[1,4] := 0.0; a[1,5] := 0.0;
   a[2,1] := -2.0; a[2,2] := 3.0; a[2,3] := 0.0;
   a[2,4] := 0.0; a[2,5] := 0.0;
   a[3,1] := 3.0; a[3,2] := 4.0; a[3,3] := 50.0;
   a[3,4] := 0.0; a[3,5] := 0.0;
   a[4,1] := -4.0; a[4,2] := 5.0; a[4,3] := -60.0;
   a[4,4] := 7.0; a[4,5] := 0.0;
   a[5,1] := -5.0; a[5,2] := 6.0; a[5,3] := -70.0;
   a[5,4] := 8.0; a[5,5] := -9.0;
   writeln('matrix:');
   FOR i := 1 TO np DO BEGIN
      FOR j := 1 TO np DO write(a[i,j]:12:2);
      writeln
   END;
   balanc(a,np);
   elmhes(a,np);
   hqr(a,np,wr,wi);
   writeln('eigenvalues:');
   writeln('real':11,'imag.':16);
   FOR i := 1 TO np DO writeln('   ',wr[i]:13,'   ',wi[i]:13)
END.
```

Chapter 12: Fourier Methods

Chapter 12 of Numerical Recipes covers Fourier transform spectral methods, particularly the transform of discretely sampled data. Central to the chapter is the fast Fourier transform (FFT). Routine `four1` *performs the FFT on a complex data array.* `twofft` *does the same transform on two real-valued data arrays (at the same time) and returns two complex-valued transforms. Finally,* `realft` *finds the Fourier transform of a single real-valued array. Two related transforms are the sine transform and the cosine transform, given by* `sinft` *and* `cosft`.

Two common uses of the Fourier transform are the convolution of data with a response function, and the computation of the correlation of two data sets. These operations are carried out by `convlv` *and* `correl` *respectively. Other applications of Fourier methods include data filtering, power spectrum estimation (*`spctrm`*, or* `evlmem` *with* `memcof`*), and linear prediction (*`predic` *with* `fixrts`*). All of these applications assume data in one dimension. For FFTs in two or more dimensions the routine* `fourn` *is supplied.*

$$\star \quad \star \quad \star \quad \star$$

Routine `four1` performs the fast Fourier transform on a complex-valued array of data points. Example program d12r1 has five tests for this transform. First, it checks the following four symmetries (where $h(t)$ is the data and $H(n)$ is the transform):

1. If $h(t)$ is real-valued and even, then $H(n) = H(N - n)$ and H is real.

2. If $h(t)$ is imaginary-valued and even, then $H(n) = H(N - n)$ and H is imaginary.

3. If $h(t)$ is real-valued and odd, then $H(n) = -H(N - n)$ and H is imaginary.

4. If $h(t)$ is imaginary-valued and odd, then $H(n) = -H(N - n)$ and H is real.

The fifth test is that if a data array is Fourier transformed twice and then inverse-transformed, the resulting array should be identical to the original.

```
PROGRAM d12r1(input,output);
(* driver for routine FOUR1 *)
(*$I MODFILE.PAS *)
CONST
   nn = 32;
   nn2 = 64; (* 2*nn *)
TYPE
   RealArrayNN2 = ARRAY [1..nn2] OF real;
VAR
   ii,i,isign: integer;
   data,dcmp: RealArrayNN2;
```

```
PROCEDURE prntft(data: RealArrayNN2; nn: integer);
VAR
   ii,mm,n: integer;
BEGIN
   writeln('n':4,'real(n)':13,'imag.(n)':13,'real(N-n)':12,'imag.(N-n)':13);
   writeln(0:4,data[1]:14:6,data[2]:12:6,data[1]:12:6,data[2]:12:6);
   mm := nn DIV 2;
   FOR ii := 1 TO mm DO BEGIN
      n := 2*ii+1;
      writeln(((n-1) DIV 2):4,data[n]:14:6,data[n+1]:12:6,
         data[2*nn+2-n]:12:6,data[2*nn+3-n]:12:6)
   END;
   writeln(' press RETURN to continue ...');
   readln
END;
(*$I FOUR1.PAS *)
BEGIN
   writeln('h(t) := real-valued even-function');
   writeln('h(n) := h(N-n) and real?');
   FOR ii := 1 TO nn DO BEGIN
      i := 2*ii-1;
      data[i] := 1.0/(sqr((i-nn-1.0)/nn)+1.0);
      data[i+1] := 0.0
   END;
   isign := 1;
   four1(data,nn,isign);
   prntft(data,nn);
   writeln('h(t) := imaginary-valued even-function');
   writeln('h(n) := h(N-n) and imaginary?');
   FOR ii := 1 TO nn DO BEGIN
      i := 2*ii-1;
      data[i+1] := 1.0/(sqr((i-nn-1.0)/nn)+1.0);
      data[i] := 0.0
   END;
   isign := 1;
   four1(data,nn,isign);
   prntft(data,nn);
   writeln('h(t) := real-valued odd-function');
   writeln('h(n) := -h(N-n) and imaginary?');
   FOR ii := 1 TO nn DO BEGIN
      i := 2*ii-1;
      data[i] := ((i-nn-1.0)/nn)/(sqr((i-nn-1.0)/nn)+1.0);
      data[i+1] := 0.0
   END;
   data[1] := 0.0;
   isign := 1;
   four1(data,nn,isign);
   prntft(data,nn);
   writeln('h(t) := imaginary-valued odd-function');
   writeln('h(n) := -h(N-n) and real?');
   FOR ii := 1 TO nn DO BEGIN
      i := 2*ii-1;
      data[i+1] := ((i-nn-1.0)/nn)/(sqr((i-nn-1.0)/nn)+1.0);
      data[i] := 0.0
   END;
   data[2] := 0.0;
   isign := 1;
```

```
      four1(data,nn,isign);
      prntft(data,nn);
(* transform, inverse-transform test *)
      FOR ii := 1 TO nn DO BEGIN
         i := 2*ii-1;
         data[i] := 1.0/(sqr(0.5*(i-nn-1.0)/nn)+1.0);
         dcmp[i] := data[i];
         data[i+1] := (0.25*(i-nn-1.0)/nn)
               *exp(-sqr(0.5*(i-nn-1.0)/nn)));
         dcmp[i+1] := data[i+1]
      END;
      isign := 1;
      four1(data,nn,isign);
      isign := -1;
      four1(data,nn,isign);
      writeln;
      writeln('double fourier transform:':33,'original data:':23);
      writeln;
      writeln('k':4,'real h(k)':14,'imag h(k)':13,
         'real h(k)':17,'imag h(k)':12);
      FOR ii := 1 TO nn DIV 2 DO BEGIN
         i := 2*ii-1;
         writeln(((i+1) DIV 2):4,dcmp[i]:14:6,dcmp[i+1]:12:6,
            data[i]/nn:17:6,data[i+1]/nn:12:6)
      END
END.
```

twofft is a routine that performs an efficient FFT of two real arrays at once by packing them into a complex array and transforming with four1. Sample program d12r2 generates two periodic data sets, out of phase with one another, and performs a transform and an inverse transform on each. It will be difficult to judge whether the transform itself gives the right answer, but if the inverse transform gets you back to the easily recognized original, you may be fairly confident that the routine works.

```
PROGRAM d12r2(input,output);
(* driver for routine TWOFFT *)
(*$I MODFILE.PAS *)
CONST
   n = 32;
   n2 = 64;    (* n2=2*n *)
   per = 8;
   pi = 3.1415926;
TYPE
   RealArrayNP = ARRAY [1..n] OF real;
   RealArray2tN = ARRAY [1..n2] OF real;
   RealArrayNN2 = RealArray2tN;
VAR
   i,isign: integer;
   data1,data2: RealArrayNP;
   fft1,fft2: RealArray2tN;
PROCEDURE prntft(data: RealArrayNN2; nn: integer);
VAR
   ii,mm,n: integer;
BEGIN
   writeln('n':4,'real(n)':13,'imag.(n)':13,'real(N-n)':12,'imag.(N-n)':13);
   writeln(0:4,data[1]:14:6,data[2]:12:6,data[1]:12:6,data[2]:12:6);
```

```
      mm := nn DIV 2;
      FOR ii := 1 TO mm DO BEGIN
         n := 2*ii+1;
         writeln(((n-1) DIV 2):4,data[n]:14:6,data[n+1]:12:6,
            data[2*nn+2-n]:12:6,data[2*nn+3-n]:12:6)
      END;
      writeln(' press return to continue ...');
      readln
   END;
(*$I FOUR1.PAS *)
(*$I TWOFFT.PAS *)
BEGIN
   FOR i := 1 TO n DO BEGIN
      data1[i] := round(cos(i*2.0*pi/per));
      data2[i] := round(sin(i*2.0*pi/per));
   END;
   twofft(data1,data2,fft1,fft2,n);
   writeln('fourier transform of first function:');
   prntft(fft1,n);
   writeln('fourier transform of second function:');
   prntft(fft2,n);
(* invert transform *)
   isign := -1;
   four1(fft1,n,isign);
   writeln('inverted transform  =  first function:');
   prntft(fft1,n);
   four1(fft2,n,isign);
   writeln('inverted transform  =  second function:');
   prntft(fft2,n)
END.
```

realft performs the Fourier transform of a single real-valued data array. Sample routine d12r3 takes this function to be sinusoidal, and allows you to choose the period. After transforming, it simply plots the magnitude of each element of the transform. If the period you choose is a power of two, the transform will be nonzero in a single bin; otherwise there will be leakage to adjacent channels. d12r3 follows every transform by an inverse transform to make sure the original function is recovered. Entering a period less than or equal to zero terminates the program.

```
PROGRAM d12r3(input,output);
(* driver for routine REALFT *)
(*$I MODFILE.PAS *)
LABEL 99;
CONST
   eps = 1.0e-3;
   np = 32;
   width = 50.0;
   pi = 3.1415926;
TYPE
   RealArrayNN2 = ARRAY [1..np] OF real;
VAR
   big,per,scal,small: real;
   i,j,n,nlim: integer;
   data,size: RealArrayNN2;
(*$I FOUR1.PAS *)
(*$I REALFT.PAS *)
```

```
BEGIN
   n := np DIV 2;
   WHILE true DO BEGIN
      writeln('Period of sinusoid in channels (2-',np:2,')');
      readln(per);
      IF per <= 0.0 THEN GOTO 99;
      FOR i := 1 TO np DO data[i] := cos(2.0*pi*(i-1)/per);
      realft(data,n,+1);
      size[1] := data[1];
      big := size[1];
      FOR i := 2 TO n DO BEGIN
         size[i] := sqrt(sqr(data[2*i-1])+sqr(data[2*i]));
         IF size[i] > big THEN
            big := size[i]
      END;
      scal := width/big;
      FOR i := 1 TO n DO BEGIN
         nlim := round(scal*size[i]+eps);
         write(i:4,' ');
         FOR j := 1 TO nlim+1 DO write('*');
         writeln
      END;
      writeln('press RETURN to continue ...');
      readln;
      realft(data,n,-1);
      big := -1.0e10;
      small := 1.0e10;
      FOR i := 1 TO np DO BEGIN
         IF data[i] < small THEN small := data[i];
         IF data[i] > big THEN big := data[i]
      END;
      scal := width/(big-small);
      FOR i := 1 TO np DO BEGIN
         nlim := round(scal*(data[i]-small)+eps);
         write(i:4,' ');
         FOR j := 1 TO nlim+1 DO write('*');
         writeln
      END
   END;
99:
END.
```

sinft performs a sine-transform of a real-valued array. The necessity for such a transform arises in solution methods for partial differential equations with certain kinds of boundary conditions (see Chapter 17). The sample program d12r4 works exactly as the previous program. Notice that in this program no distinction needs to be made between the transform and its inverse. They are identical. Entering a period less than or equal to zero terminates the program.

```
PROGRAM d12r4(input,output);
(* driver for routine SINFT *)
(*$I MODFILE.PAS *)
LABEL 99;
CONST
   eps = 1.0e-3;
   np = 16;
```

```
    width = 30.0;
    pi = 3.1415926;
TYPE
    RealArrayNN2 = ARRAY [1..np] OF real;
    RealArrayNP = RealArrayNN2;
VAR
    big,per,scal,small: real;
    i,j,nlim: integer;
    data: RealArrayNN2;
(*$I FOUR1.PAS *)
(*$I REALFT.PAS *)
(*$I SINFT.PAS *)
BEGIN
    WHILE true DO BEGIN
        writeln('period of sinusoid in channels (3-',np:2,')');
        readln(per);
        IF per <= 0.0 THEN GOTO 99;
        FOR i := 1 TO np DO data[i] := sin(2.0*pi*(i-1)/per);
        sinft(data,np);
        big := -1.0e10;
        small := 1.0e10;
        FOR i := 1 TO np DO BEGIN
            IF data[i] < small THEN  small := data[i];
            IF data[i] > big THEN  big := data[i]
        END;
        scal := width/(big-small);
        FOR i := 1 TO np DO BEGIN
            nlim := round(scal*(data[i]-small)+eps);
            write(i:4,' ');
            FOR j := 1 TO nlim+1 DO write('*');
            writeln
        END;
        writeln('press RETURN to continue ...');
        readln;
        sinft(data,np);
        big := -1.0e10;
        small := 1.0e10;
        FOR i := 1 TO np DO BEGIN
            IF data[i] < small THEN small := data[i];
            IF data[i] > big THEN big := data[i]
        END;
        scal := width/(big-small);
        FOR i := 1 TO np DO BEGIN
            nlim := round(scal*(data[i]-small)+eps);
            write(i:4,' ');
            FOR j := 1 TO nlim+1 DO write('*');
            writeln
        END
    END;
99:
END.
```

cosft is a companion procedure to sinft that does the cosine transform. It also plays a role in partial differential equation solutions. Although program d12r5 is again the same as d12r3, you will notice some difference in solutions. The cosine transform of a cosine with a period that is a power of two does not give a transform that is nonzero in a single bin. It has some small values at other frequencies. This is due to our desire to cast the transform into something that calls `realft`, and therefore works on 2^N points rather than the more natural $2^N + 1$. The sample program will prove to you, however, that the transform expressed here is invertible. Notice that, unlike the sine transform, the cosine transform is not self-inverting. Once again, entering a period less than or equal to zero terminates the program.

```pascal
PROGRAM d12r5(input,output);
(* driver for routine COSFT *)
(*$I MODFILE.PAS *)
LABEL 99;
CONST
    eps = 1.0e-3;
    np = 16;
    width = 30.0;
    pi = 3.1415926;
TYPE
    RealArrayNN2 = ARRAY [1..np] OF real;
    RealArrayNP = RealArrayNN2;
VAR
    big,per,scal,small: real;
    i,jj,j,nlim: integer;
    data: RealArrayNN2;
(*$I FOUR1.PAS *)
(*$I REALFT.PAS *)
(*$I COSFT.PAS *)
BEGIN
    WHILE true DO BEGIN
        writeln('period of cosine in channels (2-',np:2,')');
        readln(per);
        IF per <= 0.0 THEN GOTO 99;
        FOR i := 1 TO np DO data[i] := cos(2.0*pi*(i-1)/per);
        cosft(data,np,+1);
        big := -1.0e10;
        small := 1.0e10;
        FOR i := 1 TO np DO BEGIN
            IF data[i] < small THEN small := data[i];
            IF data[i] > big THEN big := data[i]
        END;
        scal := width/(big-small);
        FOR i := 1 TO np DO BEGIN
            nlim := round(scal*(data[i]-small)+eps);
            write(i:4,' ');
            FOR j := 1 TO nlim+1 DO write('*');
            writeln
        END;
        writeln('press RETURN to continue ...');
        readln;
        cosft(data,np,-1);
        big := -1.0e10;
        small := 1.0e10;
        FOR i := 1 TO np DO BEGIN
            IF data[i] < small THEN small := data[i];
            IF data[i] > big THEN big := data[i]
        END;
        scal := width/(big-small);
        FOR i := 1 TO np DO BEGIN
```

```
          nlim := round(scal*(data[i]-small)+eps);
          write(i:4,' ');
          FOR j := 1 TO nlim+1 DO write('*');
          writeln
      END
  END;
99:
END.
```

Procedure `convlv` performs the convolution of a data set with a response function using an FFT. Sample program d12r6 uses two functions that take on only the values 0.0 and 1.0. The data array `data[i]` has sixteen values, and is zero everywhere except between `i=6` and `i=10` where it is 1.0. The response function `respns[i]` has nine values and is zero except between `i=3` and `i=6` where it is 1.0. The expected value of the convolution is determined simply by flipping the response function end-to-end, moving it to the left by the desired shift, and counting how many non-zero channels of `respns` fall on non-zero channels of `data`. In this way, you should be able to verify the result from the program. The sample program, incidentally, does the calculation by this direct method for the purpose of comparison.

```
PROGRAM d12r6(input,output);
(* driver for routine CONVLV *)
(*$I MODFILE.PAS *)
CONST
    n = 16;          (* data array size *)
    m = 9;           (* response function dimension - must be odd *)
    n2 = 32;         (* n2=2*n *)
    pi = 3.14159265;
TYPE
    RealArrayNP = ARRAY [1..n] OF real;
    RealArrayN2 = ARRAY [1..n2] OF real;
    RealArray2tN = RealArrayN2;
    RealArrayNN2 = RealArrayN2;
VAR
    i,isign,j: integer;
    cmp: real;
    ans: RealArrayN2;
    data,respns,resp: RealArrayNP;
(*$I FOUR1.PAS *)
(*$I TWOFFT.PAS *)
(*$I REALFT.PAS *)
(*$I CONVLV.PAS *)
BEGIN
    FOR i := 1 TO n DO BEGIN
        data[i] := 0.0;
        IF (i >= n DIV 2 - n DIV 8) AND (i <= n DIV 2 + n DIV 8) THEN
            data[i] := 1.0
    END;
    FOR i := 1 TO m DO BEGIN
        respns[i] := 0.0;
        IF (i > 2) AND (i < 7) THEN respns[i] := 1.0;
        resp[i] := respns[i]
    END;
    isign := 1;
    convlv(data,n,resp,m,isign,ans);
(* compare with a direct convolution *)
```

```
      writeln('i':3,'CONVLV':14,'Expected':13);
      FOR i := 1 TO n DO BEGIN
         cmp := 0.0;
         FOR j := 1 TO m DIV 2 DO BEGIN
            cmp := cmp+data[((i-j-1+n) MOD n)+1]*respns[j+1];
            cmp := cmp+data[((i+j-1) MOD n)+1]*respns[m-j+1]
         END;
         cmp := cmp+data[i]*respns[1];
         writeln(i:3,ans[i]:15:6,cmp:12:6)
      END
END.
```

correl calculates the correlation function of two data sets. Sample program d12r7 defines data1[i] as an array of 64 values which are all zero except from i=25 to i=39, where they are one. data2[i] is defined in the same way. Therefore, the correlation being performed is an autocorrelation. The sample routine compares the result of the calculation as performed by correl with that found by a direct calculation. In this case the calculation may be done manually simply by successively shifting data2 with respect to data1 and counting the number of nonzero channels of the two that overlap.

```
PROGRAM d12r7(input,output);
(* driver for routine CORREL *)
(*$I MODFILE.PAS *)
CONST
   n = 64;
   n2 = 128;        (* n2=2*n *)
   pi = 3.1415927;
TYPE
   RealArrayNP = ARRAY [1..n] OF real;
   RealArray2tN = ARRAY [1..n2] OF real;
   RealArrayNN2 = RealArray2tN;
VAR
   i,j: integer;
   cmp: real;
   data1,data2: RealArrayNP;
   ans: RealArray2tN;
(*$I FOUR1.PAS *)
(*$I TWOFFT.PAS *)
(*$I REALFT.PAS *)
(*$I CORREL.PAS *)
BEGIN
   FOR i := 1 TO n DO BEGIN
      data1[i] := 0.0;
      IF (i > n DIV 2 - n DIV 8) AND (i < n DIV 2 + n DIV 8) THEN
         data1[i] := 1.0;
      data2[i] := data1[i]
   END;
   correl(data1,data2,n,ans);
(* calculate directly *)
   writeln('n':3,'CORREL':14,'direct calc.':18);
   FOR i := 0 TO 16 DO BEGIN
      cmp := 0.0;
      FOR j := 1 TO n DO
         cmp := cmp+data1[((i+j-1) MOD n)+1]*data2[j];
      writeln(i:3,ans[i+1]:15:6,cmp:15:6)
```

```
    END
END.
```

spctrm does a spectral estimate of a data set by reading it in as segments, windowing, Fourier transforming, and accumulating the power spectrum. Data segments may or may not be overlapped at the decision of the user. In sample program d12r8 the spectral data is read in from a file called spctrl.dat containing 1200 numbers and included on the *Numerical Recipes Examples Diskette*. It is analyzed first with overlap and then without. The results are tabulated side by side for comparison.

```
PROGRAM d12r8(input,output,dfile);
(* driver for routine SPCTRM *)
(*$I MODFILE.PAS *)
CONST
   m = 16;
   m4 = 64;    (* m4=4*m *)
TYPE
   RealArrayMP = ARRAY [1..m] OF real;
   RealArray4tM = ARRAY [1..m4] OF real;
   RealArrayNN2 = RealArray4tM;
VAR
   j,k: integer;
   ovrlap: boolean;
   p,q: RealArrayMP;
   dfile: text;
(*$I FOUR1.PAS *)
(*$I SPCTRM.PAS *)
BEGIN
   NROpen(dfile,'spctrl.dat');
   k := 8;
   ovrlap := true;
   spctrm(p,m,k,ovrlap);
   close(dfile);
   NROpen(dfile,'spctrl.dat');
   k := 16;
   ovrlap := false;
   spctrm(q,m,k,ovrlap);
   close(dfile);
   writeln('Spectrum of data in file SPCTRL.DAT');
   writeln(' ':13,'overlapped ',' ':5,'non-overlapped');
   FOR j := 1 TO m DO writeln(j:6,' ':5,p[j]:13,' ':5,q[j]:13)
END.
```

memcof and evlmem are used to perform spectral analysis by the maximum entropy method. memcof finds the coefficients for a model spectrum, the magnitude squared of the inverse of a polynomial series. Sample program d12r9 determines the coefficients for 1000 numbers from the file spctrl.dat and simply prints the results for comparison to the following table:

```
Coefficients for spectral estimation of spctrl.dat
  a[ 1] =    1.261539
  a[ 2] =   -0.007695
  a[ 3] =   -0.646778
  a[ 4] =   -0.280603
  a[ 5] =    0.163693
  a[ 6] =    0.347674
```

```
a[ 7] =     0.111247
a[ 8] =    -0.337141
a[ 9] =    -0.358043
a[10] =     0.378774
   a0 =     0.003511

PROGRAM d12r9(input,output,dfile);
(* driver for routine MEMCOF *)
(*$I MODFILE.PAS *)
CONST
   n = 1000;
   m = 10;
TYPE
   RealArrayNP = ARRAY [1..n] OF real;
   RealArrayMP = ARRAY [1..m] OF real;
VAR
   i: integer;
   pm: real;
   cof: RealArrayMP;
   data: RealArrayNP;
   dfile: text;
(*$I MEMCOF.PAS *)
BEGIN
   NROpen(dfile,'spctrl.dat');
   FOR i := 1 TO n DO read(dfile,data[i]);
   close(dfile);
   memcof(data,n,m,pm,cof);
   writeln('Coefficients for spectral estimation of SPCTRL.DAT');
   writeln;
   FOR i := 1 TO m DO writeln('a[',i:2,'] =',cof[i]:12:6);
   writeln;
   writeln('a0 =',pm:12:6)
END.
```

evlmem uses coefficients from memcof to generate a spectral estimate. The example d12r10 uses the same data from spctrl.dat and prints the spectral estimate. You may compare the result to:

```
Power spectrum estimate of data in spctrl.dat
     f*delta         power
     0.000000      0.026023
     0.031250      0.029266
     0.062500      0.193087
     0.093750      0.139241
     0.125000     29.915518
     0.156250      0.003878
     0.187500      0.000633
     0.218750      0.000334
     0.250000      0.000437
     0.281250      0.001331
     0.312500      0.000780
     0.343750      0.000451
     0.375000      0.000784
     0.406250      0.001381
     0.437500      0.000649
     0.468750      0.000775
     0.500000      0.001716
```

```
PROGRAM d12r10(input,output,dfile);
(* driver for routine EVLMEM *)
(*$I MODFILE.PAS *)
CONST
    n = 1000;
    m = 10;
    nfdt = 16;
TYPE
    RealArrayNP = ARRAY [1..n] OF real;
    RealArrayMP = ARRAY [1..m] OF real;
VAR
    i: integer;
    fdt,pm: real;
    cof: RealArrayMP;
    data: RealArrayNP;
    dfile: text;
(*$I MEMCOF.PAS *)
(*$I EVLMEM.PAS *)
BEGIN
    NROpen(dfile,'spctrl.dat');
    FOR i := 1 TO n DO read(dfile,data[i]);
    close(dfile);
    memcof(data,n,m,pm,cof);
    writeln('Power spectrum estimate of data in SPCTRL.DAT');
    writeln(' ':4,'f*delta',' ':7,'power');
    FOR i := 0 TO nfdt DO BEGIN
        fdt := 0.5*i/nfdt;
        writeln(fdt:12:6,evlmem(fdt,cof,m,pm):12:6)
    END
END.
```

Notice that once `memcof` has determined coefficients, we may evaluate the estimate at any intervals we wish. Notice also that we have built a spectral peak into the noisy data in `spctrl.dat`.

Linear prediction is carried out by routines `predic`, `memcof`, and `fixrts`. `memcof` produces the linear prediction coefficients from the data set. `fixrts` massages the coefficients so that all roots of the characteristic polynomial fall inside the unit circle of the complex domain, thus insuring stability of the prediction algorithm. Finally, `predic` predicts future data points based on the modified coefficients. Sample program `d12r11` demonstrates the operation of `fixrts`. The coefficients provided in the array `d[i]` are those appropriate to the polynomial $(z-1)^6 = 1$. This equation has six roots on a circle of radius one, centered at $(1.0, 0.0)$ in the complex plane. Some of these lie within the unit circle and some outside. The ones outside are moved by `fixrts` according to $z_i \rightarrow 1/z_i{}^*$. You can easily figure these out by hand and check the results. Also, the sample routine calculates $(z-1)^6$ for each of the adjusted roots, and thereby shows which have been changed and which have not.

```
PROGRAM d12r11(input,output);
(* driver for routine FIXRTS *)
(*$I MODFILE.PAS *)
CONST
    npoles = 6;
    npolp1 = 7;    (* npolp1 = npoles+1 *)
TYPE
```

```
    RealArrayNPOLES = ARRAY [1..npoles] OF real;
    Complex = RECORD
                   r,i: real
               END;
    ComplexArrayMp1 = ARRAY [1..npolp1] OF Complex;
VAR
    i,j: integer;
    dum: real;
    polish: boolean;
    d: RealArrayNPOLES;
    sixth: Complex;
    zcoef,zeros: ComplexArrayMp1;
(*$I LAGUER.PAS *)
(*$I ZROOTS.PAS *)
(*$I FIXRTS.PAS *)
BEGIN
    d[1] := 6.0; d[2] := -15.0; d[3] := 20.0;
    d[4] := -15.0; d[5] := 6.0; d[6] := 0.0;
    polish := true;
    (* finding roots of (z-1.0)^6 := 1.0 *)
    (* first write roots *)
    zcoef[npoles+1].r := 1.0;
    zcoef[npoles+1].i := 0.0;
    FOR i := npoles DOWNTO 1 DO BEGIN
        zcoef[i].r := -d[npoles+1-i];
        zcoef[i].i := 0.0
    END;
    zroots(zcoef,npoles,zeros,polish);
    writeln('Roots of (z-1.0)^6 = 1.0');
    writeln('Root':22,'(z-1.0)^6':27);
    FOR i := 1 TO npoles DO BEGIN
        sixth.r := 1.0;
        sixth.i := 0.0;
        FOR j := 1 TO 6 DO BEGIN
            dum := sixth.r;
            sixth.r := sixth.r*(zeros[i].r-1.0)-sixth.i*zeros[i].i;
            sixth.i := dum*zeros[i].i+sixth.i*(zeros[i].r-1.0)
        END;
        writeln(i:6,zeros[i].r:12:6,zeros[i].i:12:6,sixth.r:12:6,sixth.i:12:6)
    END;
    (* now fix them to lie within unit circle *)
    fixrts(d,npoles);
    (* check results *)
    zcoef[npoles+1].r := 1.0;
    zcoef[npoles+1].i := 0.0;
    FOR i := npoles DOWNTO 1 DO BEGIN
        zcoef[i].r := -d[npoles+1-i];
        zcoef[i].i := 0.0
    END;
    zroots(zcoef,npoles,zeros,polish);
    writeln;
    writeln('Roots reflected in unit circle');
    writeln('Root':22,'(z-1.0)^6':27);
    FOR i := 1 TO npoles DO BEGIN
        sixth.r := 1.0;
        sixth.i := 0.0;
        FOR j := 1 TO 6 DO BEGIN
```

```
            dum := sixth.r;
            sixth.r := sixth.r*(zeros[i].r-1.0)-sixth.i*zeros[i].i;
            sixth.i := dum*zeros[i].i+sixth.i*(zeros[i].r-1.0)
         END;
         writeln(i:6,zeros[i].r:12:6,zeros[i].i:12:6,sixth.r:12:6,sixth.i:12:6)
      END
END.
```

predic carries out the job of performing the prediction. The function chosen for
investigation in sample program d12r12 is

$$F(n) = \exp(-n/\text{npts})\sin(2\pi n/50) + \exp(-2n/\text{npts})\sin(2.2\pi n/50)$$

the sum of two sine waves of similar period and exponentially decaying amplitudes.
On the basis of 300 data points, and working with coefficients representing ten poles,
the routine predicts 20 future points. The quality of this prediction may be judged
by comparing these 20 points with the evaluations of $F(n)$ that are provided.

```
PROGRAM d12r12(input,output);
(* driver for routine PREDIC *)
(*$I MODFILE.PAS *)
CONST
   npts = 300;
   npoles = 10;
   npolp1 = 11;    (* npolp1 = npoles+1 *)
   nfut = 20;
   pi = 3.1415926;
TYPE
   RealArrayNP = ARRAY [1..npts] OF real;
   RealArrayNDATA = RealArrayNP;
   RealArrayMP = ARRAY [1..npoles] OF real;
   RealArrayNPOLES = RealArrayMP;
   RealArrayNFUT = ARRAY [1..nfut] OF real;
   Complex = RECORD
                  r,i: real
             END;
   ComplexArrayMp1 = ARRAY [1..npolp1] OF Complex;
VAR
   i: integer;
   dum: real;
   d: RealArrayMP;
   data: RealArrayNP;
   future: RealArrayNFUT;
FUNCTION f(n,npts: integer): real;
CONST
   pi = 3.1415926;
BEGIN
   f := exp(-1.0*n/npts)*sin(2.0*pi*n/50.0)
       +exp(-2.0*n/npts)*sin(2.2*pi*n/50.0)
END;
(*$I LAGUER.PAS *)
(*$I ZROOTS.PAS *)
(*$I MEMCOF.PAS *)
(*$I FIXRTS.PAS *)
(*$I PREDIC.PAS *)
BEGIN
   FOR i := 1 TO npts DO data[i] := f(i,npts);
```

```
   memcof(data,npts,npoles,dum,d);
   fixrts(d,npoles);
   predic(data,npts,d,npoles,future,nfut);
   writeln('I':6,'Actual':11,'PREDIC':12);
   FOR i := 1 TO nfut DO
      writeln(i:6,f(i+npts,npts):12:6,future[i]:12:6)
END.
```

fourn is a routine for performing N-dimensional Fourier transforms. We have used it in sample program d12r13 to transform a 3-dimensional complex data array of dimensions $4 \times 8 \times 16$. The function analyzed is not that easy to visualize, but it is very easy to calculate. The test conducted here is to perform a 3-dimensional transform and inverse transform in succession, and to compare the result with the original array. Ratios are provided for convenience.

```
PROGRAM d12r13(input,output);
(* driver for routine FOURN *)
(*$I MODFILE.PAS *)
CONST
   ndim = 3;
   ndat2 = 1024;
TYPE
   RealArrayNDAT2 = ARRAY [1..ndat2] OF real;
   IntegerArrayNDIM = ARRAY [1..ndim] OF integer;
VAR
   i,isign,j,k,l,ll,ndum: integer;
   data: RealArrayNDAT2;
   nn: IntegerArrayNDIM;
(*$I FOURN.PAS *)
BEGIN
   ndum := 2;
   FOR i := 1 TO ndim DO BEGIN
      ndum := ndum*2;
      nn[i] := ndum
   END;
   FOR k := 1 TO nn[1] DO BEGIN
      FOR j := 1 TO nn[2] DO BEGIN
         FOR i := 1 TO nn[3] DO BEGIN
            l := k+(j-1)*nn[1]+(i-1)*nn[2]*nn[1];
            ll := 2*l-1;
            data[ll] := ll;
            data[ll+1] := ll+1
         END
      END
   END;
   isign := +1;
   fourn(data,nn,ndim,isign);
   isign := -1;
   writeln('Double 3-dimensional transform');
   writeln;
   writeln('Double transf.':22,'Original data':24,'Ratio':20);
   writeln('real':10,'imag.':13,'real':12,'imag.':13,'real':11,'imag.':13);
   writeln;
   fourn(data,nn,ndim,isign);
   FOR i := 1 TO 4 DO BEGIN
      j := 2*i;
```

```
       k := 2*j;
       l := k+(j-1)*nn[1]+(i-1)*nn[2]*nn[1];
       ll := 2*l-1;
       writeln(data[ll]:12:2,data[ll+1]:12:2,ll:10,ll+1:12,
           data[ll]/ll:14:2,data[ll+1]/(ll+1):12:2)
   END;
   writeln;
   writeln('The product of transform lengths is:',nn[1]*nn[2]*nn[3]:4)
END.
```

Chapter 13: Statistical Description of Data

Chapter 13 of Numerical Recipes covers the subject of descriptive statistics, the representation of data in terms of its statistical properties, and the use of such properties to compare data sets. There are three procedures that characterize data sets. moment *returns the average, average deviation, standard deviation, variance, skewness, and kurtosis of a data array.* mdian1 *and* mdian2 *both find the median of an array. The former also sorts the array.*

Most of the remaining procedures compare data sets. ttest *compares the means of two data sets having the same variance;* tutest *does the same for two sets having different variance; and* tptest *does it for paired samples, correcting for covariance.* ftest *is a test of whether two data arrays have significantly different variance. The question of whether two distributions are different is treated by four procedures (pertaining to whether the data is binned or continuous, and whether data is compared to a model distribution or to other data). Specifically,*

1. chsone *compares binned data to a model distribution.*

2. chstwo *compares two binned data sets.*

3. ksone *compares the cumulative distribution function of an unbinned data set to a given function.*

4. kstwo *compares the cumulative distribution functions of two unbinned data sets.*

The next set of procedures tests for associations between nominal variables. cntab1 *and* cntab2 *both check for associations in a two-dimensional contingency table, the first calculating on the basis of χ^2, and the second by evaluating entropies. Linear correlation is represented by Pearson's r, or the linear correlation coefficient, which is calculated with routine* pearsn. *Alternatively, the data can be investigated with a nonparametric or rank correlation, using* spear *to find Spearman's rank correlation r_s. Kendall's τ uses rank ordering of ordinal data to test for monotonic correlations.* kendl1 *does this for two data arrays of the same size, while* kendl2 *applies it to contingency tables.*

One final routine smooft *makes no attempt to describe or compare data statistically. It seeks, instead, to smooth out the statistical fluctuations, usually for the purpose of visual presentation.*

$$\star \quad \star \quad \star \quad \star$$

Procedure **moment** calculates successive moments of a given distribution of data. The example program d13r0 creates an unusual distribution, one that has a sinusoidal distribution of values (over a half-period of the sine, so the distribution is a symmetrical peak). We have worked out the moments of such a distribution theoretically and recorded them in the program for comparison. The data is discrete and will only approximate these values.

```pascal
PROGRAM d13r0(input,output);
(* driver for routine MOMENT *)
(*$I MODFILE.PAS *)
CONST
   pi = 3.14159265;
   npts = 5000;
   nbin = 100;
   nppnb = 5100;    (* nppnb=npts+nbin *)
TYPE
   RealArrayNP = ARRAY [1..nppnb] OF real;
VAR
   adev,ave,curt,sdev,skew: real;
   vrnce,x: real;
   i,j,k,nlim: integer;
   data: RealArrayNP;
(*$I MOMENT.PAS *)
BEGIN
   i := 1;
   FOR j := 1 TO nbin DO BEGIN
      x := pi*j/nbin;
      nlim := round(sin(x)*pi/2.0*npts/nbin);
      FOR k := 1 TO nlim DO BEGIN
         data[i] := x;
         i := i+1
      END
   END;
   writeln('moments of a sinusoidal distribution');
   writeln;
   moment(data,i-1,ave,adev,sdev,vrnce,skew,curt);
   writeln('calculated':39,'expected':11);
   writeln;
   writeln('Mean :',' ':19,ave:12:4,pi/2.0:12:4);
   writeln('Average Deviation :',' ':6,adev:12:4,(pi/2.0)-1.0:12:4);
   writeln('Standard Deviation :',' ':5,sdev:12:4,0.683667:12:4);
   writeln('Variance :',' ':15,vrnce:12:4,0.467401:12:4);
   writeln('Skewness :',' ':15,skew:12:4,0.0:12:4);
   writeln('Kurtosis :',' ':15,curt:12:4,-0.806249:12:4)
END.
```

mdian1 and mdian2 both find the median of a distribution. In programs d13r1 and d13r2 we allow this distribution to be Gaussian, as produced by routine **gasdev**. This distribution should have a mean of zero and variance of one. mdian1 also sorts the data, so d13r1 prints the sorted data to show that it is done properly. Example d13r2 has nothing to show from mdian2 but the median itself, and it is checked by comparing to the result from mdian1.

```pascal
PROGRAM d13r1(input,output);
(* driver for routine MDIAN1 *)
(*$I MODFILE.PAS *)
```

```
CONST
   npts = 50;
TYPE
   RealArray55 = ARRAY [1..55] OF real;
   RealArrayNP = ARRAY [1..npts] OF real;
VAR
   Ran3Inext,Ran3Inextp: integer;
   Ran3Ma: RealArray55;
   GasdevIset: integer;
   GasdevGset: real;
   i,j,idum: integer;
   xmed: real;
   data: RealArrayNP;
(*$I RAN3.PAS *)
(*$I GASDEV.PAS *)
(*$I SORT.PAS *)
(*$I MDIAN1.PAS *)
BEGIN
   GasdevIset := 0;
   idum := -5;
   FOR i := 1 TO npts DO data[i] := gasdev(idum);
   mdian1(data,npts,xmed);
   writeln('Data drawn from a gaussian distribution');
   writeln('with zero mean and unit variance');
   writeln;
   writeln('Median of data set is',xmed:9:6);
   writeln;
   writeln('Sorted data');
   FOR i := 1 TO npts DIV 5 DO BEGIN
      FOR j := 1 TO 5 DO write(data[5*i-5+j]:12:6);
      writeln
   END
END.


PROGRAM d13r2(input,output);
(* driver for routine MDIAN2 *)
(*$I MODFILE.PAS *)
CONST
   npts = 50;
TYPE
   RealArray55 = ARRAY [1..55] OF real;
   RealArrayNP = ARRAY [1..npts] OF real;
VAR
   Ran3Inext,Ran3Inextp: integer;
   Ran3Ma: RealArray55;
   GasdevIset: integer;
   GasdevGset: real;
   i,idum: integer;
   xmed: real;
   data: RealArrayNP;
(*$I RAN3.PAS *)
(*$I GASDEV.PAS *)
(*$I SORT.PAS *)
(*$I MDIAN1.PAS *)
(*$I MDIAN2.PAS *)
BEGIN
   GasdevIset := 0;
```

```
   idum := -5;
   FOR i := 1 TO npts DO data[i] := gasdev(idum);
   mdian2(data,npts,xmed);
   writeln('Data drawn from a gaussian distribution');
   writeln('with zero mean, unit variance');
   writeln;
   writeln('median according to mdian2 is',xmed:9:6);
   mdian1(data,npts,xmed);
   writeln('median according to mdian1 is',xmed:9:6)
END.
```

Student's *t*-test is a test of two data sets for significantly different means. It is applied by d13r3 to two Gaussian data sets data1 and data2 that are generated by gasdev. data2 is originally given an artificial shift of its mean to the right of that of data1, by nshft/2 units of eps. Then data1 is successively shifted nshft times to the right by eps and compared to data2 by ttest. At about step nshft/2, the two distributions should superpose and indicate populations with the same mean. Notice that the two populations have the same variance (i.e. 1.0), as required by ttest.

```
PROGRAM d13r3(input,output);
(* driver for routine TTEST *)
(* generate gaussian distributed data *)
(*$I MODFILE.PAS *)
CONST
   npts = 1024;
   mpts = 512;
   eps = 0.03;
   nshft = 10;
TYPE
   RealArray55 = ARRAY [1..55] OF real;
   RealArrayN12 = ARRAY [1..npts] OF real;    (* max. of npts and mpts *)
   RealArrayNP = RealArrayN12;
VAR
   Ran3Inext,Ran3Inextp: integer;
   Ran3Ma: RealArray55;
   GasdevIset: integer;
   GasdevGset: real;
   data1,data2: RealArrayN12;
   i,idum,j: integer;
   prob,t: real;
(*$I AVEVAR.PAS *)
(*$I GAMMLN.PAS *)
(*$I BETACF.PAS *)
(*$I BETAI.PAS *)
(*$I RAN3.PAS *)
(*$I GASDEV.PAS *)
(*$I TTEST.PAS *)
BEGIN
   GasdevIset := 0;
   idum := -11;
   FOR i := 1 TO npts DO data1[i] := gasdev(idum);
   FOR i := 1 TO mpts DO data2[i] := (nshft DIV 2)*eps+gasdev(idum);
   writeln('shift':6,'t':8,'probability':16);
   FOR i := 1 TO nshft+1 DO BEGIN
      ttest(data1,npts,data2,mpts,t,prob);
      writeln((i-1)*eps:6:2,t:10:2,prob:10:2);
```

```
    FOR j := 1 TO npts DO data1[j] := data1[j]+eps
  END
END.
```

avevar is an auxiliary routine for ttest. It finds the average and variance of a data set. Sample program d13r4 generates a series of Gaussian distributions for i=1,..,11, and gives each a shift of $(i-1)$eps and a variance of i^2. This progression allows you easily to check the operation of avevar "by eye".

```
PROGRAM d13r4(input,output);
(* driver for routine AVEVAR *)
(*$I MODFILE.PAS *)
CONST
  npts = 1000;
  eps = 0.1;
TYPE
  RealArray55 = ARRAY [1..55] OF real;
  RealArrayNP = ARRAY [1..npts] OF real;
VAR
  GasdevIset: integer;
  GasdevGset: real;
  Ran3Inext,Ran3Inextp: integer;
  Ran3Ma: RealArray55;
  i,idum,j: integer;
  ave,shift,vrnce: real;
  data: RealArrayNP;
(*$I RAN3.PAS *)
(*$I GASDEV.PAS *)
(*$I AVEVAR.PAS *)
BEGIN
(* generate gaussian distributed data *)
  GasdevIset := 0;
  idum := -5;
  writeln('shift':9,'average':11,'variance':12);
  FOR i := 1 TO 11 DO BEGIN
    shift := (i-1)*eps;
    FOR j := 1 TO npts DO
      data[j] := shift+i*gasdev(idum);
    avevar(data,npts,ave,vrnce);
    writeln(shift:8:2,ave:11:2,vrnce:12:2)
  END
END.
```

tutest also does Student's t-test, but applies to the comparison of means of two distributions with different variance. The example d13r5 employs the comparison used on ttest but gives the two distributions data1 and data2 variances of 1.0 and 4.0 respectively.

```
PROGRAM d13r5(input,output);
(* driver for routine TUTEST *)
(*$I MODFILE.PAS *)
CONST
  npts = 3000;
  mpts = 600;
  eps = 0.03;
  var1 = 1.0;
  var2 = 4.0;
```

```
    nshft = 10;
TYPE
    RealArray55 = ARRAY [1..55] OF real;
    RealArrayN12 = ARRAY [1..npts] OF real;    (* max of npts and mpts *)
    RealArrayNP = RealArrayN12;
VAR
    Ran3Inext,Ran3Inextp: integer;
    Ran3Ma: RealArray55;
    GasdevIset: integer;
    GasdevGset: real;
    fctr1,fctr2,prob,t: real;
    i,idum,j: integer;
    data1,data2: RealArrayN12;
(*$I AVEVAR.PAS *)
(*$I GAMMLN.PAS *)
(*$I BETACF.PAS *)
(*$I BETAI.PAS *)
(*$I RAN3.PAS *)
(*$I GASDEV.PAS *)
(*$I TUTEST.PAS *)
BEGIN
(* generate two gaussian distributions of different variance *)
    GasdevIset := 0;
    idum := -1773;
    fctr1 := sqrt(var1);
    FOR i := 1 TO npts DO data1[i] := fctr1*gasdev(idum);
    fctr2 := sqrt(var2);
    FOR i := 1 TO mpts DO data2[i] := (nshft DIV 2)*eps+fctr2*gasdev(idum);
    writeln;
    writeln('Distribution #1 : variance = ',var1:6:2);
    writeln('Distribution #2 : variance = ',var2:6:2);
    writeln;
    writeln('shift':7,'t':8,'probability':16);
    FOR i := 1 TO nshft+1 DO BEGIN
        tutest(data1,npts,data2,mpts,t,prob);
        writeln((i-1)*eps:6:2,t:10:2,prob:11:2);
        FOR j := 1 TO npts DO data1[j] := data1[j]+eps
    END
END.
```

tptest goes a step further, and compares two distributions not only having different variances, but also perhaps having point by point correlations. The example d13r6 creates two situations, one with correlated and one with uncorrelated distributions. It does this by way of three data sets. data1 is a simple Gaussian distribution of zero mean and unit variance. data2 is data1 plus some additional Gaussian fluctuations of smaller amplitude. data3 is similar to data2 but generated with independent calls to gasdev so that its fluctuations ought not to have any correlation with those of data1. data1 is then given an offset with respect to the others and they are successively shifted as in previous routines. At each step of the shift tptest was applied. Our results are given below:

Shift	Correlated:		Uncorrelated:	
	t	Probability	t	Probability
0.01	2.9551	0.0033	0.6411	0.5218
0.02	2.2163	0.0271	0.4808	0.6309
0.03	1.4775	0.1402	0.3205	0.7487

0.04	0.7388	0.4604	0.1603	0.8727
0.05	0.0000	1.0000	0.0000	1.0000
0.06	-0.7388	0.4604	-0.1603	0.8727
0.07	-1.4775	0.1402	-0.3205	0.7487
0.08	-2.2163	0.0271	-0.4808	0.6309
0.09	-2.9551	0.0033	-0.6411	0.5218
0.10	-3.6939	0.0002	-0.8014	0.4233

```pascal
PROGRAM d13r6(input,output);
(* driver for routine TPTEST *)
(* compare two correlated distributions vs. two *)
(* uncorrelated distributions *)
(*$I MODFILE.PAS *)
CONST
   npts = 500;
   eps = 0.01;
   nshft = 10;
   anoise = 0.3;
TYPE
   RealArray55 = ARRAY [1..55] OF real;
   RealArrayNP = ARRAY [1..npts] OF real;
VAR
   GasdevIset: integer;
   GasdevGset: real;
   Ran3Inext,Ran3Inextp: integer;
   Ran3Ma: RealArray55;
   ave1,ave2,ave3,gauss: real;
   offset,prob1,prob2,shift,t1,t2: real;
   var1,var2,var3: real;
   i,idum,j: integer;
   data1,data2,data3: RealArrayNP;
(*$I RAN3.PAS *)
(*$I GASDEV.PAS *)
(*$I GAMMLN.PAS *)
(*$I BETACF.PAS *)
(*$I BETAI.PAS *)
(*$I AVEVAR.PAS *)
(*$I TPTEST.PAS *)
BEGIN
   GasdevIset := 0;
   idum := -5;
   writeln('Correlated:':29,'Uncorrelated:':30);
   writeln('Shift':7,'t':11,'Probability':17,'t':11,'Probability':17);
   offset := (nshft DIV 2)*eps;
   FOR j := 1 TO npts DO BEGIN
      gauss := gasdev(idum);
      data1[j] := gauss;
      data2[j] := gauss+anoise*gasdev(idum);
      data3[j] := gasdev(idum)+anoise*gasdev(idum)
   END;
   avevar(data1,npts,ave1,var1);
   avevar(data2,npts,ave2,var2);
   avevar(data3,npts,ave3,var3);
   FOR j := 1 TO npts DO BEGIN
      data1[j] := data1[j]-ave1+offset;
      data2[j] := data2[j]-ave2;
      data3[j] := data3[j]-ave3
```

```
      END;
      FOR i := 1 TO nshft DO BEGIN
         shift := i*eps;
         FOR j := 1 TO npts DO BEGIN
            data2[j] := data2[j]+eps;
            data3[j] := data3[j]+eps
         END;
         tptest(data1,data2,npts,t1,prob1);
         tptest(data1,data3,npts,t2,prob2);
         writeln(shift:6:2,t1:14:4,prob1:12:4,t2:16:4,prob2:12:4)
      END
END.
```

The *F*-test (procedure `ftest`) is a test for differing variances between two distributions. For demonstration purposes, sample program `d13r7` generates a Gaussian distribution `data1` of unit variance. The values of a second Gaussian distribution `data2` are then set by multiplying `data1` by a series of values `factor` which takes its variance from 1.0 to 1.4 in ten equal steps. The effect of this on the *F*-test can be evaluated from the probabilities `prob`.

```
PROGRAM d13r7(input,output);
(* driver for routine FTEST *)
(*$I MODFILE.PAS *)
CONST
   npts = 1000;
   mpts = 500;
   eps = 0.04;
   nval = 10;
TYPE
   RealArray55 = ARRAY [1..55] OF real;
   RealArrayN12 = ARRAY [1..npts] OF real;   (* max of npts and mpts *)
   RealArrayNP = RealArrayN12;
VAR
   Ran3Inext,Ran3Inextp: integer;
   Ran3Ma: RealArray55;
   GasdevIset: integer;
   GasdevGset: real;
   f,factor,prob,vrnce: real;
   i,idum,j: integer;
   data1,data2: RealArrayN12;
(*$I RAN3.PAS *)
(*$I GASDEV.PAS *)
(*$I GAMMLN.PAS *)
(*$I BETACF.PAS *)
(*$I BETAI.PAS *)
(*$I AVEVAR.PAS *)
(*$I FTEST.PAS *)
BEGIN
(* generate two gaussian distributions with
different variances *)
   GasdevIset := 0;
   idum := -144;
   writeln;
   writeln('Variance 1 = ':16,1.0:5:2);
   writeln('Variance 2':13,'Ratio':11,'Probability':16);
   FOR i := 1 TO nval+1 DO BEGIN
```

```
      FOR j := 1 TO npts DO
         data1[j] := gasdev(idum);
      vrnce := 1.0+(i-1)*eps;
      factor := sqrt(vrnce);
      FOR j := 1 TO mpts DO
         data2[j] := factor*gasdev(idum);
      ftest(data1,npts,data2,mpts,f,prob);
      writeln(vrnce:11:4,f:13:4,prob:13:4)
   END
END.
```

chsone and chstwo compare two distributions on the basis of a χ^2 test to see if they are different. chsone, specifically, compares a data distribution to an expected distribution. Sample program d13r8 generates an exponential distribution bins[i] of data using routine expdev. It then creates an array ebins[i] which is the expected result (a smooth exponential decay in the absence of statistical fluctuations). ebins and bins are compared by chsone to give χ^2 and a probability that they represent the same distribution.

```
PROGRAM d13r8(input,output);
(* driver for routine CHSONE *)
(*$I MODFILE.PAS *)
CONST
   nbins = 10;
   npts = 2000;
TYPE
   RealArray55 = ARRAY [1..55] OF real;
   RealArrayNBINS = ARRAY [0..nbins] OF real;
VAR
   Ran3Inext,Ran3Inextp: integer;
   Ran3Ma: RealArray55;
   chsq,df,prob,x: real;
   i,ibin,idum,j: integer;
   bins,ebins: RealArrayNBINS;
(*$I GAMMLN.PAS *)
(*$I GCF.PAS *)
(*$I GSER.PAS *)
(*$I GAMMQ.PAS *)
(*$I RAN3.PAS *)
(*$I EXPDEV.PAS *)
(*$I CHSONE.PAS *)
BEGIN
   idum := -15;
   FOR j := 1 TO nbins DO bins[j] := 0.0;
   FOR i := 1 TO npts DO BEGIN
      x := expdev(idum);
      ibin := trunc(x*nbins/3.0)+1;
      IF ibin <= nbins THEN
         bins[ibin] := bins[ibin]+1.0
   END;
   FOR i := 1 TO nbins DO
      ebins[i] := 3.0*npts/nbins*exp(-3.0*(i-0.5)/nbins);
   chsone(bins,ebins,nbins,-1,df,chsq,prob);
   writeln('expected':15,'observed':15);
   FOR i := 1 TO nbins DO
      writeln(ebins[i]:14:2,bins[i]:15:2);
```

```
     writeln;
     writeln('chi-squared:':19,chsq:10:4);
     writeln('probability:':19,prob:10:4)
END.
```

chstwo compares two binned distributions bins1 and bins2, again using a χ^2 test. Sample program d13r9 prepares these distributions both in the same way. Each is composed of 2000 random numbers, drawn from an exponential deviate, and placed into 10 bins. The two data sets are then analyzed by chstwo to calculate χ^2 and probability prob.

```
PROGRAM d13r9(input,output);
(* driver for routine CHSTWO *)
(*$I MODFILE.PAS *)
CONST
   nbins = 10;
   npts = 2000;
TYPE
   RealArray55 = ARRAY [1..55] OF real;
   RealArrayNBINS = ARRAY [1..nbins] OF real;
VAR
   Ran3Inext,Ran3Inextp: integer;
   Ran3Ma: RealArray55;
   chsq,df,prob,x: real;
   i,ibin,idum,j: integer;
   bins1,bins2: RealArrayNBINS;
(*$I GAMMLN.PAS *)
(*$I GSER.PAS *)
(*$I GCF.PAS *)
(*$I GAMMQ.PAS *)
(*$I RAN3.PAS *)
(*$I EXPDEV.PAS *)
(*$I CHSTWO.PAS *)
BEGIN
   idum := -18;
   FOR j := 1 TO nbins DO BEGIN
      bins1[j] := 0.0;
      bins2[j] := 0.0
   END;
   FOR i := 1 TO npts DO BEGIN
      x := expdev(idum);
      ibin := trunc(x*nbins/3.0)+1;
      IF ibin <= nbins THEN
         bins1[ibin] := bins1[ibin]+1.0;
      x := expdev(idum);
      ibin := trunc(x*nbins/3.0)+1;
      IF ibin <= nbins THEN
         bins2[ibin] := bins2[ibin]+1.0
   END;
   chstwo(bins1,bins2,nbins,-1,df,chsq,prob);
   writeln;
   writeln('dataset 1':15,'dataset 2':15);
   FOR i := 1 TO nbins DO
      writeln(bins1[i]:13:2,bins2[i]:15:2);
   writeln;
   writeln('chi-squared:':18,chsq:12:4);
   writeln('probability:':18,prob:12:4)
```

END.

The Kolmogorov-Smirnov test used in **ksone** and **kstwo** applies to unbinned distributions with a single independent variable. **ksone** uses the K-S criterion to compare a single data set to an expected distribution, and **kstwo** uses it to compare two data sets. Sample program d13r10 creates data sets with Gaussian distributions and with stepwise increasing variance, and compares their cumulative distribution function to the expected result for a Gaussian distribution of unit variance. This result is the error function and is generated by routine **erfcc**. Increasing variance in the test distribution should reduce the likelihood that it was drawn from the same distribution represented by the comparison function.

```pascal
PROGRAM d13r10(input,output);
(* driver for routine KSONE *)
(*$I MODFILE.PAS *)
CONST
   npts = 1000;
   eps = 0.1;
TYPE
   RealArray55 = ARRAY [1..55] OF real;
   RealArrayNP = ARRAY [1..npts] OF real;
VAR
   Ran3Inext,Ran3Inextp: integer;
   Ran3Ma: RealArray55;
   GasdevIset: integer;
   GasdevGset: real;
   i,idum,j: integer;
   d,factr,prob,varnce: real;
   data: RealArrayNP;
(*$I SORT.PAS *)
(*$I PROBKS.PAS *)
(*$I RAN3.PAS *)
(*$I GASDEV.PAS *)
(*$I ERFCC.PAS *)
FUNCTION func(x: real): real;
VAR
   y: real;
BEGIN
   y := x/sqrt(2.0);
   func := 1.0 - erfcc(y)
END;
(*$I KSONE.PAS *)
BEGIN
   GasdevIset := 0;
   idum := -5;
   writeln('variance ratio':19,'k-s statistic':16,'probability':15);
   writeln;
   FOR i := 1 TO 11 DO BEGIN
      varnce := 1.0+(i-1)*eps;
      factr := sqrt(varnce);
      FOR j := 1 TO npts DO
         data[j] := factr*abs(gasdev(idum));
      ksone(data,npts,d,prob);
      writeln(varnce:16:6,d:16:6,prob:16:6)
   END
END.
```

kstwo compares the cumulative distribution functions of two unbinned data sets, data1 and data2. In sample program d13r11, they are both Gaussian distributions, but data2 is given a stepwise increase of variance. In other respects, d13r11 is like d13r10.

```
PROGRAM d13r11(input,output);
(* driver for routine KSTWO *)
(*$I MODFILE.PAS *)
CONST
   n1 = 500;
   n2 = 100;
   eps = 0.2;
TYPE
   RealArray55 = ARRAY [1..55] OF real;
   RealArrayN12 = ARRAY [1..n1] OF real;    (* max of n1 and n2 *)
   RealArrayNP = RealArrayN12;
VAR
   Ran3Inext,Ran3Inextp: integer;
   Ran3Ma: RealArray55;
   GasdevIset: integer;
   GasdevGset: real;
   i,idum,j: integer;
   d,factr,prob,varnce: real;
   data1,data2: RealArrayN12;
(*$I PROBKS.PAS *)
(*$I SORT.PAS *)
(*$I RAN3.PAS *)
(*$I GASDEV.PAS *)
(*$I KSTWO.PAS *)
BEGIN
   GasdevIset := 0;
   idum := -1357;
   FOR i := 1 TO n1 DO data1[i] := gasdev(idum);
   writeln('variance ratio':18,'k-s statistic':15,'probability':14);
   idum := -2468;
   FOR i := 1 TO 11 DO BEGIN
      varnce := 1.0+(i-1)*eps;
      factr := sqrt(varnce);
      FOR j := 1 TO n2 DO
         data2[j] := factr*gasdev(idum);
      kstwo(data1,n1,data2,n2,d,prob);
      writeln(varnce:15:6,d:15:6,prob:15:6)
   END
END.
```

probks is an auxiliary routine for ksone and kstwo which calculates the function $Q_{ks}(\lambda)$ used to evaluate the probability that the two distributions being compared are the same. There is no independent means of producing this function, so in sample program d13r12 we have chosen simply to graph it. Our output is reproduced below.

```
PROGRAM d13r12(input,output);
(* driver for routine PROBKS *)
(*$I MODFILE.PAS *)
CONST
   npts = 20;
   eps = 0.1;
   iscal = 40;
```

```
TYPE
    CharArrayISCAL = PACKED ARRAY [1..iscal] OF char;
VAR
    alam,aval: real;
    i,j: integer;
    txt: CharArrayISCAL;
(*$I PROBKS.PAS *)
BEGIN
    writeln('probability function for kolmogorov-smirnov statistic');
    writeln;
    writeln('lambda':7,'value:':10,'graph:':13);
    FOR i := 1 TO npts DO BEGIN
        alam := i*eps;
        aval := probks(alam);
        FOR j := 1 TO iscal DO
            IF j <= round(iscal*aval) THEN txt[j] := '*'
            ELSE txt[j] := ' ';
        writeln(alam:8:6,aval:10:6,' ':5,txt)
    END
END.
```

```
Probability func. for Kolmogorov-Smirnov statistic
  Lambda:     Value:     Graph:
  0.100000    1.000000   ******************************************
  0.200000    1.000000   ******************************************
  0.300000    0.999991   ******************************************
  0.400000    0.997192   ******************************************
  0.500000    0.963945   *****************************************
  0.600000    0.864283   **********************************
  0.700000    0.711235   ****************************
  0.800000    0.544142   **********************
  0.900000    0.392731   ****************
  1.000000    0.270000   ***********
  1.100000    0.177718   *******
  1.200000    0.112250   ****
  1.300000    0.068092   ***
  1.400000    0.039682   **
  1.500000    0.022218   *
  1.600000    0.011952
  1.700000    0.006177
  1.800000    0.003068
  1.900000    0.001464
  2.000000    0.000671
```

Procedure cntab1 analyzes a two-dimensional contingency table and returns several parameters describing any association between its nominal variables. Sample program d13r13 supplies a table from a file table1.dat which is listed in the Appendix to this chapter. The table shows the rate of certain accidents, tabulated on a monthly basis. These data are listed, as well as their statistical properties, by d13r13. We found the results to be:

Chi-squared	5026.30
Degrees of Freedom	88.00
Probability	.0000
Cramer-V	.0772
Contingency Coeff.	.2134

```pascal
PROGRAM d13r13(input,output,dfile);
(* driver for routine CNTAB1 *)
(* contingency table in file TABLE1.DAT *)
(*$I MODFILE.PAS *)
CONST
   ndat = 9;
   nmon = 12;
   ni = ndat;
   nj = nmon;
TYPE
   IntegerArrayNIbyNJ = ARRAY [1..ni,1..nj] OF integer;
   RealArrayNI = ARRAY [1..ni] OF real;
   RealArrayNJ = ARRAY [1..nj] OF real;
   StrArray15 = string[15];
   StrArray5 = string[5];
   StrArray64 = string[64];
VAR
   ccc,chisq,cramrv,df,prob: real;
   i,j: integer;
   nmbr: IntegerArrayNIbyNJ;
   fate: ARRAY [1..ndat] OF StrArray15;
   mon: ARRAY [1..nmon] OF StrArray5;
   txt: StrArray64;
   dfile: text;
(*$I GAMMLN.PAS *)
(*$I GCF.PAS *)
(*$I GSER.PAS *)
(*$I GAMMQ.PAS *)
(*$I CNTAB1.PAS *)
BEGIN
   NROpen(dfile,'table1.dat');
   readln(dfile);
   readln(dfile,txt);
   read(dfile,fate[1]);
   FOR i := 1 TO 12 DO read(dfile,mon[i]);
   readln(dfile);
   readln(dfile);
   FOR i := 1 TO ndat DO BEGIN
      read(dfile,fate[i]);
      FOR j := 1 TO 12 DO read(dfile,nmbr[i,j]);
      readln(dfile)
   END;
   close(dfile);
   writeln;
   writeln(txt);
   writeln;
   write(' ':15);
   FOR i := 1 TO 12 DO write(mon[i]:5);
   writeln;
   FOR i := 1 TO ndat DO BEGIN
      write(fate[i]);
      FOR j := 1 TO 12 DO write(nmbr[i,j]:5);
      writeln
   END;
   cntab1(nmbr,ndat,nmon,chisq,df,prob,cramrv,ccc);
   writeln;
   writeln('chi-squared        ',chisq:20:2);
```

```
      writeln('degrees of freedom',df:20:2);
      writeln('probability     ',prob:20:4);
      writeln('cramer-v        ',cramrv:20:4);
      writeln('contingency coeff.',ccc:20:4)
END.
```

The test looks for any association between accidents and the months in which they occur. table1.dat clearly shows some. Drownings, for example, happen mostly in the summer. cntab2 carries out a similar analysis on table1.dat but measures associations on the basis of entropy. Sample program d13r14 prints out the following entropies for the table:

Entropy of Table	4.0368
Entropy of x-distribution	1.5781
Entropy of y-distribution	2.4820
Entropy of y given x	2.4588
Entropy of x given y	1.5548
Dependency of y on x	.0094
Dependency of x on y	.0147
Symmetrical dependency	.0114

```
PROGRAM d13r14(input,output,dfile);
(* driver for routine CNTAB2 *)
(* contingency table in file TABLE1.DAT *)
(*$I MODFILE.PAS *)
CONST
   ni = 9;
   nmon = 12;
   nj = nmon;
TYPE
   IntegerArrayNIbyNJ = ARRAY [1..ni,1..nmon] OF integer;
   RealArrayNI = ARRAY [1..ni] OF real;
   RealArrayNJ = ARRAY [1..nj] OF real;
   StrArray5 = string[5];
   StrArray15 = string[15];
   StrArray64 = string[64];
VAR
   h,hx,hxgy,hy,hygx: real;
   uxgy,uxy,uygx: real;
   i,j: integer;
   nmbr: IntegerArrayNIbyNJ;
   fate: ARRAY [1..ni] OF StrArray15;
   mon: ARRAY [1..nmon] OF StrArray5;
   txt: StrArray64;
   dfile: text;
(*$I CNTAB2.PAS *)
BEGIN
   NROpen(dfile,'table1.dat');
   readln(dfile);
   readln(dfile,txt);
   read(dfile,fate[1]);
   FOR i := 1 TO 12 DO read(dfile,mon[i]);
   readln(dfile);
   readln(dfile);
   FOR i := 1 TO ni DO BEGIN
      read(dfile,fate[i]);
      FOR j := 1 TO 12 DO read(dfile,nmbr[i,j]);
```

```
      readln(dfile)
   END;
   close(dfile);
   writeln;
   writeln(txt);
   writeln;
   write(' ':15);
   FOR i := 1 TO 12 DO write(mon[i]:5);
   writeln;
   FOR i := 1 TO ni DO BEGIN
      write(fate[i]);
      FOR j := 1 TO 12 DO write(nmbr[i,j]:5);
      writeln
   END;
   cntab2(nmbr,ni,nmon,h,hx,hy,hygx,hxgy,uygx,uxgy,uxy);
   writeln;
   writeln('entropy of table           ',h:10:4);
   writeln('entropy of x-distribution  ',hx:10:4);
   writeln('entropy of y-distribution  ',hy:10:4);
   writeln('entropy of y given x       ',hygx:10:4);
   writeln('entropy of x given y       ',hxgy:10:4);
   writeln('dependency of y on x       ',uygx:10:4);
   writeln('dependency of x on y       ',uxgy:10:4);
   writeln('symmetrical dependency     ',uxy:10:4)
END.
```

The dependencies of x on y and y on x indicate the degree to which the type of accident can be predicted by knowing the month, or vice-versa.

pearsn makes an examination of two ordinal or continuous variables to find linear correlations. It returns a linear correlation coefficient r, a probability of correlation prob, and Fisher's z. Sample program d13r15 sets up data pairs in arrays dose and spore which show hypothetical data for the spore count from plants exposed to various levels of γ-rays. The results of applying pearsn to this data set are compared with the correct results by the program.

```
PROGRAM d13r15(input,output);
(* driver for routine PEARSN *)
(*$I MODFILE.PAS *)
CONST
   n = 10;
TYPE
   RealArrayNP = ARRAY [1..n] OF real;
VAR
   prob,r,z: real;
   i: integer;
   dose,spore: RealArrayNP;
(*$I GAMMLN.PAS *)
(*$I BETACF.PAS *)
(*$I BETAI.PAS *)
(*$I PEARSN.PAS *)
BEGIN
   dose[1] := 56.1; dose[2] := 64.1; dose[3] := 70.0;
   dose[4] := 66.6; dose[5] := 82.0; dose[6] := 91.3;
   dose[7] := 90.0; dose[8] := 99.7; dose[9] := 115.3;
   dose[10] := 110.0;
   spore[1] := 0.11; spore[2] := 0.40; spore[3] := 0.37;
```

```
    spore[4] := 0.48; spore[5] := 0.75; spore[6] := 0.66;
    spore[7] := 0.71; spore[8] := 1.20; spore[9] := 1.01;
    spore[10] := 0.95;
    writeln;
    writeln('Effect of Gamma Rays on Man-in-the-Moon Marigolds');
    writeln('Count Rate (cpm)':16,'Pollen Index':23);
    FOR i := 1 TO 10 DO writeln(dose[i]:10:2,spore[i]:25:2);
    pearsn(dose,spore,10,r,prob,z);
    writeln;
    writeln('PEARSN':30,'Expected':16);
    writeln('Corr. Coeff.',' ':8,r:13,' ',0.9069586:13);
    writeln('Probability',' ':9,prob:13,' ',0.2926505e-3:13);
    writeln('Fisher''s z',' ':10,z:13,' ',1.510110:13)
END.
```

Rank order correlation may be done with spear to compare two distributions data1 and data2 for correlation. Correlations are reported both in terms of d, the sum-squared difference in ranks, and rs, Spearman's rank correlation parameter. Sample program d13r16 applies the calculation to the data in table table2.dat (see Appendix) which shows the solar flux incident on various cities during different months of the year. It then checks for correlations between columns of the table, considering each column as a separate data set. In this fashion it looks for correlations between the July solar flux and that of other months. The probability of such correlations are shown by probd and probrs. Our results are:

Correlation of sampled U.S. solar radiation (July with other months)

month	d	st. dev.	probd	spearman-r	probrs
jul	0.00	-4.358899	0.000013	1.000000	0.000000
aug	122.00	-3.958458	0.000075	0.908133	0.000000
sep	218.00	-3.643896	0.000269	0.835967	0.000004
oct	384.00	-3.098494	0.001945	0.710843	0.000443
nov	390.50	-3.077642	0.002086	0.706060	0.000503
dec	622.00	-2.318075	0.020445	0.531803	0.015806
jan	644.50	-2.244251	0.024816	0.514866	0.020181
feb	483.50	-2.772503	0.005563	0.636056	0.002573
mar	497.00	-2.728208	0.006368	0.625894	0.003158
apr	405.50	-3.027925	0.002462	0.694654	0.000677
may	264.00	-3.492371	0.000479	0.801205	0.000022
jun	121.50	-3.960099	0.000075	0.908509	0.000000

```
PROGRAM d13r16(input,output,dfile);
(* driver for routine SPEAR *)
(*$I MODFILE.PAS *)
CONST
    ndat = 20;
    nmon = 12;
    n = ndat;
TYPE
    StrArray64 = string[64];
    StrArray15 = string[15];
    StrArray4 = string[4];
    RealArrayNDATbyNMON = ARRAY [1..ndat,1..nmon] OF real;
    RealArrayNP = ARRAY [1..n] OF real;
VAR
    d,probd,probrs,rs,zd: real;
    i,j: integer;
```

```
    ave,data1,data2,zlat: RealArrayNP;
    rays: RealArrayNDATbyNMON;
    city: ARRAY [1..ndat] OF StrArray15;
    mon: ARRAY [1..nmon] OF StrArray4;
    txt: StrArray64;
    dfile: text;
(*$I SORT2.PAS *)
(*$I ERFCC.PAS *)
(*$I GAMMLN.PAS *)
(*$I BETACF.PAS *)
(*$I BETAI.PAS *)
(*$I SPEAR.PAS *)
BEGIN
    NROpen(dfile,'table2.dat');
    readln(dfile);
    readln(dfile,txt);
    read(dfile,city[1]);
    FOR i := 1 TO nmon DO read(dfile,mon[i]);
    readln(dfile);
    readln(dfile);
    FOR i := 1 TO ndat DO BEGIN
        read(dfile,city[i]);
        FOR j := 1 TO nmon DO read(dfile,rays[i,j]);
        read(dfile,ave[i]);
        read(dfile,zlat[i]);
        readln(dfile)
    END;
    close(dfile);
    writeln(txt);
    write(' ':15);
    FOR i := 1 TO 12 DO write(mon[i]:4);
    writeln;
    FOR i := 1 TO ndat DO BEGIN
        write(city[i]:15);
        FOR j := 1 TO 12 DO write(round(rays[i,j]):4);
        writeln
    END;
(* check temperature correlations between different months *)
    writeln;
    writeln('Are sunny summer places also sunny winter places?');
    writeln('Check correlation of sampled U.S. solar radiation');
    writeln('(july with other months)');
    writeln;
    writeln('month','d':9,'st. dev.':14,'probd':11,
            'spearman-r':15,'probrs':10);
    FOR i := 1 TO ndat DO data1[i] := rays[i,1];
    FOR j := 1 TO 12 DO BEGIN
        FOR i := 1 TO ndat DO data2[i] := rays[i,j];
        spear(data1,data2,ndat,d,zd,probd,rs,probrs);
        writeln(mon[j],d:12:2,zd:12:6,probd:12:6,
                rs:13:6,probrs:12:6)
    END
END.
```

kend11 and kend12 test for monotonic correlations of ordinal data. They differ
in that kend11 compares two data sets of the same rank, while kend12 operates

on a contingency table. Sample program d13r18, for example, uses **kendl1** to look for pair correlations in three of our random number routines. That is to say, it tests for randomness by seeing if two consecutive numbers from the generator have a monotonic correlation. It uses the random number generators **ran1**, **ran2**, and **ran3**, one at a time, to generate 200 pairs of random numbers each. Then **kendl1** tests for correlation of the pairs, and a chart is made showing Kendall's τ, the standard deviation from the null hypotheses, and the probability. For a better test of the generators, you may wish to increase the number of pairs **ndat**. It would also be a good idea to see how your result depends on the value of the seed **idum**.

```
PROGRAM d13r18(input,output,infile);
(* driver for routine KENDL1 *)
(* look for correlations in ran1, ran2 and ran3 *)
(*$I MODFILE.PAS *)
CONST
   ndat = 200;
TYPE
   RealArrayNP = ARRAY [1..ndat] OF real;
   CharArray4 = PACKED ARRAY [1..4] OF char;
   RealArray55 = ARRAY [1..55] OF real;
   RealArray97 = ARRAY [1..97] OF real;
   IntegerArray97 = ARRAY [1..97] OF longint;
VAR
   infile: text;
   Ran1Ix1,Ran1Ix2,Ran1Ix3: longint;
   Ran1R: RealArray97;
   iduml,Ran2Iy: longint;
   Ran2Ir: IntegerArray97;
   Ran3Inext,Ran3Inextp: integer;
   Ran3Ma: RealArray55;
   i,idum,j: integer;
   prob,tau,z: real;
   data1,data2: RealArrayNP;
   txt: ARRAY [1..3] OF CharArray4;
(*$I RAN1.PAS *)
(*$I RAN2.PAS *)
(*$I RAN3.PAS *)
(*$I ERFCC.PAS *)
(*$I KENDL1.PAS *)
BEGIN
   txt[1] := 'RAN1'; txt[2] := 'RAN2'; txt[3] := 'RAN3';
   writeln;
   writeln('Pair correlations of RAN1, RAN2 and RAN3');
   writeln;
   writeln('Program':9,'Kendall tau':17,'Std. Dev.':16,'Probability':18);
   FOR i := 1 TO 3 DO BEGIN
      idum := -1357;
      iduml := -1357;
      FOR j := 1 TO ndat DO BEGIN
         IF i = 1 THEN BEGIN
            data1[j] := ran1(idum);
            data2[j] := ran1(idum)
         END ELSE IF i = 2 THEN BEGIN
            data1[j] := ran2(iduml);
            data2[j] := ran2(iduml)
         END ELSE IF i = 3 THEN BEGIN
```

```
          data1[j] := ran3(idum);
          data2[j] := ran3(idum)
        END
      END;
      kendl1(data1,data2,ndat,tau,z,prob);
      writeln(txt[i]:8,tau:17:6,z:17:6,prob:17:6)
    END
END.
```

Sample program d13r19, for procedure kendl2, prepares a contingency table based on the routines irbit1 and irbit2. You may recall that these routines generate random binary sequences. The program checks the sequences by breaking them into groups of three bits. Each group is treated as a three-bit binary number. Two consecutive groups then act as indices into an 8×8 contingency table that records how many times each possible sequence of six bits (two groups) occurs. For each random bit generator, ndat=1000 samples are taken. Then the contingency table tab[k,l] is analyzed by kendl2 to find Kendall's τ, the standard deviation, and the probability. Notice that Kendall's τ can only be applied when both variables are ordinal (here, the numbers 0 to 7), and that the test is specifically for monotonic correlations. In this case we are actually testing whether the larger 3-bit binary numbers tend to be followed by others of their own kind. Within the program, we have expressed this roughly as a test of whether ones or zeros tend to come in groups more than they should.

```
PROGRAM d13r19(input,output);
(* driver for routine KENDL2 *)
(* look for 'ones-after-zeros' in irbit1 and irbit2 sequences *)
(*$I MODFILE.PAS *)
CONST
   ndat = 1000;
   ip = 8;
   jp = 8;
TYPE
   RealArrayIPbyJP = ARRAY [1..ip,1..jp] OF real;
   CharArray3 = PACKED ARRAY [1..3] OF char;
VAR
   ifunc,iseed,i,j,k,l,m,n,twoton: integer;
   prob,tau,z: real;
   tab: RealArrayIPbyJP;
   txt: ARRAY [1..8] OF CharArray3;
(*$I IRBIT1.PAS *)
(*$I IRBIT2.PAS *)
(*$I ERFCC.PAS *)
(*$I KENDL2.PAS *)
BEGIN
   txt[1] := '000'; txt[2] := '001';
   txt[3] := '010'; txt[4] := '011';
   txt[5] := '100'; txt[6] := '101';
   txt[7] := '110'; txt[8] := '111';
   i := ip;
   j := jp;
   writeln('Are ones followed by zeros and vice-versa?');
   FOR ifunc := 1 TO 2 DO BEGIN
      iseed := 2468;
      IF ifunc = 1 THEN
```

```
         writeln('test of irbit1:')
      ELSE
         writeln('test of irbit2:');
      FOR k := 1 TO i DO
         FOR l := 1 TO j DO tab[k,l] := 0.0;
      FOR m := 1 TO ndat DO BEGIN
         k := 1;
         twoton := 1;
         FOR n := 0 TO 2 DO BEGIN
            IF ifunc = 1 THEN
               k := k+irbit1(iseed)*twoton
            ELSE
               k := k+irbit2(iseed)*twoton;
            twoton := 2*twoton
         END;
         l := 1;
         twoton := 1;
         FOR n := 0 TO 2 DO BEGIN
            IF ifunc = 1 THEN
               l := l+irbit1(iseed)*twoton
            ELSE
               l := l+irbit2(iseed)*twoton;
            twoton := 2*twoton
         END;
         tab[k,l] := tab[k,l]+1.0
      END;
      kendl2(tab,i,j,tau,z,prob);
      write(' ':4);
      FOR n := 1 TO 8 DO write(txt[n]:6);
      writeln;
      FOR n := 1 TO 8 DO BEGIN
         write(txt[n]:3);
         FOR m := 1 TO 8 DO write(round(tab[n,m]):6);
         writeln
      END;
      writeln;
      writeln('kendall tau':17,'std. dev.':14,'probability':16);
      writeln(tau:15:6,z:15:6,prob:15:6);
      writeln
   END
END.
```

smooft is a procedure for smoothing data. This is not a mathematically valuable procedure since it always reduces the information content of the data. However, it is a satisfactory tool for data presentation, as it may help to make evident important features of the data. Sample program d13r20 prepares an artificial data set y[i] with a broad maximum. It then adds noise from a Gaussian deviate. Subsequently the data is plotted three times; first the original data, then following each of two consecutive applications of smooft. You will notice that the second use of smooft is almost entirely ineffectual, but the first makes a significant change in the presentational quality of the graph.

```
PROGRAM d13r20(input,output);
(* driver for routine SMOOFT *)
(*$I MODFILE.PAS *)
CONST
```

```
   n = 100;
   hash = 0.05;
   scale = 100.0;
   pts = 10.0;
   m = 258;   (* 2 + first integral power of 2 that is greater or
                    equal to (n+2*pts) *)
TYPE
   RealArray55 = ARRAY [1..55] of real;
   RealArrayMP = ARRAY [1..m] OF real;
   RealArrayNN2 = RealArrayMP;
   CharArray50 = PACKED ARRAY [1..50] OF char;
VAR
   Ran3Inext,Ran3Inextp: integer;
   Ran3Ma: RealArray55;
   GasdevIset: integer;
   GasdevGset: real;
   bar: real;
   i,ii,idum,j,k,nstp: integer;
   y: RealArrayMP;
   txt: CharArray50;
(*$I RAN3.PAS *)
(*$I GASDEV.PAS *)
(*$I FOUR1.PAS *)
(*$I REALFT.PAS *)
(*$I SMOOFT.PAS *)
BEGIN
   GasdevIset := 0;
   idum := -7;
   FOR i := 1 TO n DO BEGIN
      y[i] := 3.0*i/n*exp(-3.0*i/n);
      y[i] := y[i]+hash*gasdev(idum)
   END;
   FOR k := 1 TO 3 DO BEGIN
      nstp := n DIV 20;
      writeln('data':8,'graph':11);
      FOR i := 1 TO 20 DO BEGIN
         ii := nstp*(i-1)+1;
         FOR j := 1 TO 50 DO txt[j] := ' ';
         bar := round(scale*y[ii]);
         FOR j := 1 TO 50 DO
            IF j <= bar THEN txt[j] := '*';
         writeln(y[ii]:10:6,' ':4,txt)
      END;
      writeln(' press return to smooth ...');
      readln;
      smooft(y,n,pts)
   END
END.
```

Appendix

File table1.dat:

Accidental Deaths by Month and Type (1979)

Month:	jan	feb	mar	apr	may	jun	jul	aug	sep	oct	nov	dec
Motor Vehicle	3298	3304	4241	4291	4594	4710	4914	4942	4861	4914	4563	4892
Falls	1150	1034	1089	1126	1142	1100	1112	1099	1114	1079	999	1181
Drowning	180	190	370	530	800	1130	1320	990	580	320	250	212
Fires	874	768	630	516	385	324	277	272	271	381	533	760
Choking	299	264	258	247	273	269	251	269	271	279	297	266
Fire-arms	168	142	122	140	153	142	147	160	162	172	266	230
Poisons	298	277	346	263	253	239	268	228	240	260	252	241
Gas-poison	267	193	144	127	70	63	55	53	60	118	150	172
Other	1264	1234	1172	1220	1547	1339	1419	1453	1359	1308	1264	1246

File table2.dat:

Average solar radiation (watts/square meter) for selected cities

Month:	jul	aug	sep	oct	nov	dec	jan	feb	mar	apr	may	jun	ave	lat
Atlanta, GA	257	246	201	166	30	102	106	140	184	236	258	271	192	34.0
Barrow, AK	208	123	56	20	0	0	0	18	87	184	248	256	100	71.0
Bismark, ND	296	251	185	132	78	60	76	121	170	217	267	284	178	47.0
Boise, ID	324	275	221	152	88	60	69	113	164	235	284	309	191	43.5
Boston, MA	240	206	165	115	70	58	67	96	142	176	228	242	150	42.5
Caribou, ME	246	218	161	102	53	51	66	111	178	194	229	232	153	47.0
Cleveland, OH	267	239	182	127	68	56	60	87	151	182	253	271	162	41.5
Dodge City, KS	311	287	239	184	138	113	123	153	202	256	275	315	216	38.0
El Paso, TX	324	309	278	224	178	151	160	209	266	317	346	353	260	32.0
Fresno, CA	323	293	243	182	117	77	90	143	212	264	308	337	216	37.0
Greensboro, NC	263	235	197	156	118	95	97	134	171	227	257	273	185	36.0
Honolulu, HI	305	293	271	245	208	176	175	200	234	262	300	297	247	21.0
Little Rock, AR	270	250	214	167	118	91	96	127	173	220	256	272	188	35.0
Miami, FL	260	246	216	188	171	154	166	201	238	263	267	257	219	26.0
New York, NY	251	238	175	127	77	62	71	102	151	183	220	255	159	41.0
Omaha, NE	275	252	192	142	96	80	99	134	172	224	248	272	182	21.0
Rapid City, SD	288	262	208	152	99	76	90	135	193	235	259	287	190	44.0
Seattle, WA	242	209	150	84	44	29	34	60	118	174	216	228	132	47.5
Tucson, AZ	304	286	281	216	172	144	151	195	264	322	358	343	253	41.0
Washington, DC	267	190	196	145	75	64	101	124	153	182	215	247	163	39.0

Chapter 14: Modeling of Data

Chapter 14 of Numerical Recipes deals with the fitting of a model function to a set of data, in order to summarize the data in terms of a few model parameters. Both traditional least-squares fitting and robust fitting are considered. Fits to a straight line are carried out by routine fit. *More general linear least-squares fits are handled by* lfit *and* covsrt. *(Remember that the term "linear" here refers not to a linear dependence of the fitting function on its argument, but rather to a linear dependence of the function on its fitting parameters.) In cases where* lfit *fails, owing probably to near degeneracy of some basis functions, the answer may still be found using* svdfit *and* svdvar. *In fact, these are generally recommended in preference to* lfit *because they never (?) fail. For nonlinear least-squares fits, the Levenberg-Marquardt method is discussed, and is implemented in* mrqmin, *which makes use also of* covsrt.

Robust estimation is discussed in several forms, and illustrated by routine medfit *which fits a straight line to data points based on the criterion of least absolute deviations rather than least-squared deviations.*

⋆   ⋆   ⋆   ⋆

Routine fit fits a set of N data points (x[i],y[i]), with standard deviations sig[i], to the linear model $y = A + Bx$. It uses χ^2 as the criterion for goodness-of-fit. To demonstrate fit, we generate some noisy data in sample program d14r1. For npt values of i we take $x = 0.1i$ and $y = 1 - 2x$ plus some values drawn from a Gaussian distribution to represent noise. Then we make two calls to fit, first performing the fit without allowance for standard deviations sig[i], and then with such allowance. Since sig[i] has been set to the constant value spread, it should not affect the resulting parameter values. The values output from this routine are:

Ignoring standard deviation:

```
a  =    1.043799       uncertainty: 0.103852
b  =   -2.011959       uncertainty: 0.017854
chi-squared:        26.029572
goodness-of-fit:     1.000000
```

Including standard deviation:

```
a  =    1.043799       uncertainty: 0.100755
b  =   -2.011959       uncertainty: 0.017321
chi-squared:       104.118286
goodness-of-fit:     0.317164
```

```
PROGRAM d14r1(input,output);
(* driver for routine FIT *)
(*$I MODFILE.PAS *)
CONST
   npt = 100;
   spread = 0.5;
TYPE
   RealArray55 = ARRAY [1..55] OF real;
   RealArrayNDATA = ARRAY [1..npt] OF real;
VAR
   GasdevIset: integer;
   GasdevGset: real;
   Ran3Inext,Ran3Inextp: integer;
   Ran3Ma: RealArray55;
   a,b,chi2,q,siga,sigb: real;
   i,idum,mwt: integer;
   x,y,sig: RealArrayNDATA;
(*$I GAMMLN.PAS *)
(*$I GSER.PAS *)
(*$I GCF.PAS *)
(*$I GAMMQ.PAS *)
(*$I RAN3.PAS *)
(*$I GASDEV.PAS *)
(*$I FIT.PAS *)
BEGIN
   GasdevIset := 0;
   idum := -117;
   FOR i := 1 TO npt DO BEGIN
      x[i] := 0.1*i;
      y[i] := -2.0*x[i]+1.0+spread*gasdev(idum);
      sig[i] := spread
   END;
   FOR mwt := 0 TO 1 DO BEGIN
      fit(x,y,npt,sig,mwt,a,b,siga,sigb,chi2,q);
      writeln;
      IF mwt = 0 THEN
         writeln('ignoring standard deviations')
      ELSE
         writeln('including standard deviation');
      writeln(' ':5,'a   = ',a:9:6,' ':6,'uncertainty:',siga:9:6);
      writeln(' ':5,'b   = ',b:9:6,' ':6,'uncertainty:',sigb:9:6);
      writeln(' ':5,'chi-squared: ',chi2:14:6);
      writeln(' ':5,'goodness-of-fit: ',q:10:6)
   END
END.
```

lfit carries out the same sort of fit but this time does a linear least-squares fit to a more general function. In sample program d14r2 the chosen function is a linear sum of powers of x, generated by procedure funcs. For convenience in checking the result we have generated data according to $y = 1 + 2x + 3x^2 + \cdots$. This series is truncated depending on the choice of nterm, and some Gaussian noise is added to simulate realistic data. The sig[i] are taken as constant errors. lfit is called three times to fit the same data. The first time lista[i] is set to i, so that the fitted parameters should be returned in the order a[1] $\approx$ 1.0, a[2] $\approx$ 2.0, a[3] $\approx$ 3.0. Then, as a test of the lista feature, which determines which parameters are to be fit and in which order, the array lista[i] is reversed. Finally, the fit is restricted to odd-numbered parameters, while even-numbered parameters are fixed. In this case the elements of the covariance matrix associated with fixed parameters should be zero. In **d14r2**, we have set **nterm=3** to fit a quadratic. You may wish to try something larger.

```pascal
PROGRAM d14r2(input,output);
(* driver for routine LFIT *)
(*$I MODFILE.PAS *)
CONST
    npt = 100;
    spread = 0.1;
    nterm = 3;
TYPE
    RealArray55 = ARRAY [1..55] OF real;
    RealArrayMAbyMA = ARRAY [1..nterm,1..nterm] OF real;
    IntegerArrayNP = ARRAY [1..nterm] OF integer;
    RealArrayNPbyNP = RealArrayMAbyMA;
    RealArrayNPbyMP = ARRAY [1..nterm,1..1] OF real;
    IntegerArrayMFIT = ARRAY [1..nterm] OF integer;
    RealArrayNDATA = ARRAY [1..npt] OF real;
    RealArrayMA = ARRAY [1..nterm] OF real;
VAR
    GasdevIset: integer;
    GasdevGset: real;
    Ran3Inext,Ran3Inextp: integer;
    Ran3Ma: RealArray55;
    chisq: real;
    i,ii,idum,j,mfit: integer;
    lista: IntegerArrayMFIT;
    a: RealArrayMA;
    covar: RealArrayMAbyMA;
    x,y,sig: RealArrayNDATA;
PROCEDURE funcs(x: real; VAR afunc: RealArrayMA; ma: integer);
(* Programs using FUNCS must define the type
TYPE
    RealArrayMA = ARRAY [1..ma] OF real;
in the main routine. *)
VAR
    i: integer;
BEGIN
    afunc[1] := 1.0;
    FOR i := 2 TO ma DO afunc[i] := x*afunc[i-1]
END;
(*$I RAN3.PAS *)
(*$I GASDEV.PAS *)
(*$I GAUSSJ.PAS *)
(*$I COVSRT.PAS *)
(*$I LFIT.PAS *)
BEGIN
    GasdevIset := 0;
    idum := -911;
    FOR i := 1 TO npt DO BEGIN
        x[i] := 0.1*i;
        y[i] := nterm;
        FOR j := nterm-1 DOWNTO 1 DO
            y[i] := j+y[i]*x[i];
        y[i] := y[i]+spread*gasdev(idum);
```

```
         sig[i] := spread
      END;
      mfit := nterm;
      FOR i := 1 TO mfit DO lista[i] := i;
      lfit(x,y,sig,npt,a,nterm,lista,mfit,covar,chisq);
      writeln;
      writeln('parameter':9,'uncertainty':23);
      FOR i := 1 TO nterm DO
         writeln('a[':4,i:1,'] = ',a[i]:8:6,sqrt(covar[i,i]):12:6);
      writeln('chi-squared = ',chisq:12);
      writeln('full covariance matrix');
      FOR i := 1 TO nterm DO BEGIN
         FOR j := 1 TO nterm DO write(covar[i,j]:12);
         writeln
      END;
      writeln;
      writeln('press RETURN to continue...');
      readln;
(* now test the LISTA feature *)
      FOR i := 1 TO nterm DO lista[i] := nterm+1-i;
      lfit(x,y,sig,npt,a,nterm,lista,mfit,covar,chisq);
      writeln('parameter':9,'uncertainty':23);
      FOR i := 1 TO nterm DO
         writeln('a[':4,i:1,'] = ',a[i]:8:6,sqrt(covar[i,i]):12:6);
      writeln('chi-squared = ',chisq:12);
      writeln('full covariance matrix');
      FOR i := 1 TO nterm DO BEGIN
         FOR j := 1 TO nterm DO write(covar[i,j]:12);
         writeln
      END;
      writeln;
      writeln('press RETURN to continue...');
      readln;
(* now check results of restricting fit parameters *)
      ii := 1;
      FOR i := 1 TO nterm DO
         IF odd(i) THEN BEGIN
            lista[ii] := i;
            ii := ii+1
         END;
      mfit := ii-1;
      lfit(x,y,sig,npt,a,nterm,lista,mfit,covar,chisq);
      writeln('parameter':9,'uncertainty':23);
      FOR i := 1 TO nterm DO
         writeln('a[':4,i:1,'] = ',a[i]:8:6,sqrt(covar[i,i]):12:6);
      writeln('chi-squared = ',chisq:12);
      writeln('full covariance matrix');
      FOR i := 1 TO nterm DO BEGIN
         FOR j := 1 TO nterm DO write(covar[i,j]:12);
         writeln
      END;
      writeln
END.
```

covsrt is used in conjunction with lfit (and later with the routine svdfit) to redistribute the covariance matrix covar so that it represents the true order of coefficients, rather than the order in which they were fit. In sample routine d14r3 an artificial 10×10 covariance matrix covar[i,j] is created, which is all zeros except for the upper left 5×5 section, for which the elements are covar[i,j]=i+j-1. Then three tests are performed.

1. By setting lista[i] $= 2i$ for $i = 1, \ldots, 5$ and mfit=5, we spread the elements so that alternate elements are zero.

2. By taking lista[i] $=$ mfit$+1-i$ for $i = 1, \ldots, 5$ we put the elements in reverse order, but leave them in an upper left-hand block.

3. With lista[i] $= 12 - 2i$ for $i = 1, \ldots, 5$ we both spread and reverse the elements.

```
PROGRAM d14r3(input,output);
(* driver for routine COVSRT *)
(*$I MODFILE.PAS *)
CONST
   ma = 10;
   mfit = 5;
TYPE
   RealArrayMAbyMA = ARRAY [1..ma,1..ma] OF real;
   IntegerArrayMFIT = ARRAY [1..mfit] OF integer;
VAR
   i,j: integer;
   covar: RealArrayMAbyMA;
   lista: IntegerArrayMFIT;
(*$I COVSRT.PAS *)
BEGIN
   FOR i := 1 TO ma DO BEGIN
      FOR j := 1 TO ma DO BEGIN
         covar[i,j] := 0.0;
         IF (i <= 5) AND (j <= 5) THEN covar[i,j] := i+j-1
      END
   END;
   writeln;
   writeln('original matrix');
   FOR i := 1 TO ma DO BEGIN
      FOR j := 1 TO ma DO write(covar[i,j]:4:1);
      writeln
   END;
   writeln(' press RETURN to continue...');
   readln;
(* test 1 - spread by 2 *)
   writeln;
   writeln('test #1 - spread by two');
   FOR i := 1 TO mfit DO lista[i] := 2*i;
   covsrt(covar,ma,lista,mfit);
   FOR i := 1 TO ma DO BEGIN
      FOR j := 1 TO ma DO write(covar[i,j]:4:1);
      writeln
   END;
   writeln(' press RETURN to continue...');
   readln;
(* test 2 - reverse *)
   writeln;
   writeln('test #2 - reverse');
```

```
      FOR i := 1 TO ma DO BEGIN
         FOR j := 1 TO ma DO BEGIN
            covar[i,j] := 0.0;
            IF (i <= 5) AND (j <= 5) THEN covar[i,j] := i+j-1
         END
      END;
      FOR i := 1 TO mfit DO lista[i] := mfit+1-i;
      covsrt(covar,ma,lista,mfit);
      FOR i := 1 TO ma DO BEGIN
         FOR j := 1 TO ma DO write(covar[i,j]:4:1);
         writeln
      END;
      writeln(' press RETURN to continue...');
      readln;
(* test 3 - spread and reverse *)
      writeln;
      writeln('test #3 - spread and reverse');
      FOR i := 1 TO ma DO BEGIN
         FOR j := 1 TO ma DO BEGIN
            covar[i,j] := 0.0;
            IF (i <= 5) AND (j <= 5) THEN covar[i,j] := i+j-1
         END
      END;
      FOR i := 1 TO mfit DO lista[i] := ma+2-2*i;
      covsrt(covar,ma,lista,mfit);
      FOR i := 1 TO ma DO BEGIN
         FOR j := 1 TO ma DO write(covar[i,j]:4:1);
         writeln
      END
END.
```

Routine svdfit is recommended in preference to lfit for performing linear least-squares fits. The sample program d14r4 puts svdfit to work on the data generated according to

$$F(x) = 1 + 2x + 3x^2 + 4x^3 + 5x^4 + \text{Gaussian noise.}$$

This data is fit to a five-term polynomial sum. sig[i], the measurement fluctuation in y, is taken to be constant. For the polynomial fit, the resulting coefficients should clearly have the values a[i] $\approx$ i.

```
PROGRAM d14r4(input,output);
(* driver for routine SVDFIT *)
(* polynomial fit *)
(*$I MODFILE.PAS *)
CONST
   npt = 100;
   spread = 0.02;
   npol = 5;
   mp = npt;
   np = npol;
TYPE
   RealArray55 = ARRAY [1..55] OF real;
   RealArrayNDATA = ARRAY [1..npt] OF real;
   RealArrayMA = ARRAY [1..npol] OF real;
   RealArrayNPbyNP = ARRAY [1..np,1..np] OF real;
   RealArrayMPbyNP = ARRAY [1..mp,1..np] OF real;
```

```
      RealArrayNP = RealArrayMA;
      RealArrayMP = RealArrayNDATA;
      RealArrayMAbyMA = RealArrayNPbyNP;
VAR
      Ran3Inext,Ran3Inextp: integer;
      Ran3Ma: RealArray55;
      GasdevIset: integer;
      GasdevGset: real;
      chisq: real;
      i,idum: integer;
      x,y,sig: RealArrayNDATA;
      a: RealArrayMA;
      cvm: RealArrayMAbyMA;
      u: RealArrayMPbyNP;
      v: RealArrayNPbyNP;
      w: RealArrayNP;
(*$I RAN3.PAS *)
(*$I GASDEV.PAS *)
(*$I SVDCMP.PAS *)
(*$I SVBKSB.PAS *)
(*$I SVDVAR.PAS *)
PROCEDURE func(x: real; VAR p: RealArrayMA; ma: integer);
(* This is essentially FPOLY renamed. *)
VAR
      j: integer;
BEGIN
      p[1] := 1.0;
      FOR j := 2 TO ma DO p[j] := p[j-1]*x
END;
(*$I SVDFIT.PAS *)
BEGIN
      GasdevIset := 0;
      idum := -911;
      FOR i := 1 TO npt DO BEGIN
         x[i] := 0.02*i;
         y[i] := 1.0+x[i]*(2.0+x[i]*(3.0+x[i]*(4.0+x[i]*5.0)));
         y[i] := y[i]*(1.0+spread*gasdev(idum));
         sig[i] := y[i]*spread
         END;
      svdfit(x,y,sig,npt,a,npol,u,v,w,chisq);
      svdvar(v,npol,w,cvm);
      writeln;
      writeln('Polynomial fit:');
      FOR i := 1 TO npol DO
         writeln(a[i]:12:6,'  +-',sqrt(cvm[i,i]):10:6);
      writeln('Chi-squared',chisq:12:6)
END.
```

svdvar is used with svdfit to evaluate the covariance matrix cvm of a fit with ma parameters. In program d14r5, we provide input vector w and array v for this routine, and then calculate the covariance matrix cvm determined from them. We have also done the calculation by hand and recorded the correct results in array tru for comparison.

```
PROGRAM d14r5(input,output);
(* driver for routine SVDVAR *)
(*$I MODFILE.PAS *)
```

```
CONST
   np = 6;
   ma = 3;
   ncvm = ma;
TYPE
   RealArrayNPbyNP = ARRAY [1..np,1..np] OF real;
   RealArrayNP = ARRAY [1..np] OF real;
   RealArrayMAbyMA = ARRAY [1..ma,1..ma] OF real;
VAR
   i,j: integer;
   v: RealArrayNPbyNP;
   w: RealArrayNP;
   cvm,tru: RealArrayMAbyMA;
(*$I SVDVAR.PAS *)
BEGIN
   w[1] := 0.0; w[2] := 1.0; w[3] := 2.0;
   w[4] := 3.0; w[5] := 4.0; w[6] := 5.0;
   v[1,1] := 1.0; v[1,2] := 1.0; v[1,3] := 1.0;
   v[1,4] := 1.0; v[1,5] := 1.0; v[1,6] := 1.0;
   v[2,1] := 2.0; v[2,2] := 2.0; v[2,3] := 2.0;
   v[2,4] := 2.0; v[2,5] := 2.0; v[2,6] := 2.0;
   v[3,1] := 3.0; v[3,2] := 3.0; v[3,3] := 3.0;
   v[3,4] := 3.0; v[3,5] := 3.0; v[3,6] := 3.0;
   v[4,1] := 4.0; v[4,2] := 4.0; v[4,3] := 4.0;
   v[4,4] := 4.0; v[4,5] := 4.0; v[4,6] := 4.0;
   v[5,1] := 5.0; v[5,2] := 5.0; v[5,3] := 5.0;
   v[5,4] := 5.0; v[5,5] := 5.0; v[5,6] := 5.0;
   v[6,1] := 6.0; v[6,2] := 6.0; v[6,3] := 6.0;
   v[6,4] := 6.0; v[6,5] := 6.0; v[6,6] := 6.0;
   tru[1,1] := 1.25; tru[1,2] := 2.5; tru[1,3] := 3.75;
   tru[2,1] := 2.5; tru[2,2] := 5.0; tru[2,3] := 7.5;
   tru[3,1] := 3.75; tru[3,2] := 7.5; tru[3,3] := 11.25;
   writeln;
   writeln('matrix v');
   FOR i := 1 TO np DO BEGIN
      FOR j := 1 TO np DO write(v[i,j]:12:6);
      writeln
   END;
   writeln;
   writeln('vector w');
   FOR i := 1 TO np DO write(w[i]:12:6);
   writeln;
   svdvar(v,ma,w,cvm);
   writeln;
   writeln('covariance matrix from svdvar');
   FOR i := 1 TO ma DO BEGIN
      FOR j := 1 TO ma DO write(cvm[i,j]:12:6);
      writeln
   END;
   writeln;
   writeln('expected covariance matrix');
   FOR i := 1 TO ma DO BEGIN
      FOR j := 1 TO ma DO write(tru[i,j]:12:6);
      writeln
   END
END.
```

Routines `fpoly` and `fleg` are used with sample program d14r4 to generate the powers of x and the Legendre polynomials, respectively. In the case of `fpoly`, sample program d14r6 is used to list the powers of x generated by `fpoly` so that they may be checked "by eye". For `fleg`, the generated polynomials in program d14r7 are compared to values from routine `plgndr`.

```pascal
PROGRAM d14r6(input,output);
(* driver for routine FPOLY *)
(*$I MODFILE.PAS *)
CONST
   nval = 15;
   dx = 0.1;
   npoly = 5;
TYPE
   RealArrayNP = ARRAY [1..npoly] OF real;
VAR
   i,j: integer;
   x: real;
   afunc: RealArrayNP;
(*$I FPOLY.PAS *)
BEGIN
   writeln;
   writeln('powers of x':38);
   writeln('x':8,'x**0':11,'x**1':10,'x**2':10,'x**3':10,'x**4':10);
   FOR i := 1 TO nval DO BEGIN
      x := i*dx;
      fpoly(x,afunc,npoly);
      write(x:10:4);
      FOR j := 1 TO npoly DO write(afunc[j]:10:4);
      writeln
   END
END.

PROGRAM d14r7(input,output);
(* driver for routine FLEG *)
(*$I MODFILE.PAS *)
CONST
   nval = 5;
   dx = 0.2;
   npoly = 5;
TYPE
   RealArrayNL = ARRAY [1..npoly] OF real;
VAR
   i,j: integer;
   x: real;
   afunc: RealArrayNL;
(*$I PLGNDR.PAS *)
(*$I FLEG.PAS *)
BEGIN
   writeln;
   writeln('legendre polynomials':43);
   writeln('n=1':9,'n=2':10,'n=3':10,'n=4':10,'n=5':10);
   FOR i := 1 TO nval DO BEGIN
      x := i*dx;
      fleg(x,afunc,npoly);
      writeln('x   := ',x:6:2);
      FOR j := 1 TO npoly DO write(afunc[j]:10:4);
```

```
        writeln(' routine FLEG');
        FOR j := 1 TO npoly DO write(plgndr(j-1,0,x):10:4);
        writeln(' routine PLGNDR');
        writeln
    END
END.
```

The procedure mrqmin is used to perform nonlinear least-squares fits with the Levenberg-Marquardt method. The artificial data used to try it in sample program d14r8 is computed as the sum of two Gaussians plus noise:

$$y[i] = a[1]\exp\{-[(x[i]-a[2])/a[3]]^2\}$$
$$+ a[4]\exp\{-[(x[i]-a[5])/a[6]]^2\} + \text{noise}.$$

The a[i] are assigned at the beginning of the program, as are the initial guesses gues[i] for these parameters to be used in initiating the fit. Also initialized for the fit are lista[i]=i for i=1,..,mfit to specify that all six of the parameters are to be fit. On the first call to mrqmin, alamda=-1 to initialize. Then a loop is entered in which mrqmin is iterated while testing successive values of chi-squared chisq. When chisq changes by less than 0.1 on two consecutive iterations, the fit is considered complete, and mrqmin is called one final time with alamda=0.0 so that array covar will return the covariance matrix. Uncertainties are derived from the square roots of the diagonal elements of covar. Expected results for the parameters are, of course, the values used to generate the "data" in the first place. The procedure fgauss, here renamed funcs, is used to generate the y[i].

```
PROGRAM D14R8(input,output);
(* driver for routine MRQMIN *)
(*$I MODFILE.PAS *)
CONST
    npt = 100;
    ma = 6;
    spread = 0.001;
TYPE
    RealArray55 = ARRAY [1..55] OF real;
    RealArrayMA = ARRAY [1..ma] OF real;
    IntegerArrayMFIT = ARRAY [1..ma] OF integer;
    RealArrayMAbyMA = ARRAY [1..ma,1..ma] OF real;
    RealArrayMAby1 = ARRAY [1..ma,1..1] OF real;
    IntegerArrayNP = IntegerArrayMFIT;
    RealArrayNPbyNP = RealArrayMAbyMA;
    RealArrayNPbyMP = RealArrayMAby1;
    RealArrayNDATA = ARRAY [1..npt] OF real;
VAR
    MrqminOchisq: real;
    MrqminBeta: RealArrayMA;
    Ran3Inext,Ran3Inextp: integer;
    Ran3Ma: RealArray55;
    GasdevIset: integer;
    GasdevGset: real;
    alamda,chisq,ochisq: real;
    i,idum,itst,j,jj,k,mfit: integer;
    x,y,sig: RealArrayNDATA;
    lista: IntegerArrayMFIT;
    a,gues: RealArrayMA;
```

```
   covar,alpha: RealArrayMAbyMA;
(*$I COVSRT.PAS *)
(*$I RAN3.PAS *)
(*$I GASDEV.PAS *)
PROCEDURE funcs(x: real; VAR a: RealArrayMA; VAR y: real;
      VAR dyda: RealArrayMA; ma: integer);
VAR
   i,ii: integer;
   fac,ex,arg: real;
BEGIN
   y := 0.0;
   FOR ii := 1 TO ma DIV 3 DO BEGIN
      i := 3*ii-2;
      arg := (x-a[i+1])/a[i+2];
      ex := exp(-sqr(arg));
      fac := a[i]*ex*2.0*arg;
      y := y+a[i]*ex;
      dyda[i] := ex;
      dyda[i+1] := fac/a[i+2];
      dyda[i+2] := fac*arg/a[i+2]
   END
END;
(*$I GAUSSJ.PAS *)
(*$I MRQMIN.PAS *)
BEGIN
   GasdevIset := 0;
   a[1] := 5.0; a[2] := 2.0; a[3] := 3.0;
   a[4] := 2.0; a[5] := 5.0; a[6] := 3.0;
   gues[1] := 4.5; gues[2] := 2.2; gues[3] := 2.8;
   gues[4] := 2.5; gues[5] := 4.9; gues[6] := 2.8;
   idum := -911;
   FOR i := 1 TO 100 DO BEGIN
      x[i] := 0.1*i;
      y[i] := 0.0;
      FOR jj := 1 TO 2 DO BEGIN
         j := 3*jj-2;
         y[i] := y[i]+a[j]*exp(-sqr((x[i]-a[j+1])/a[j+2]))
      END;
      y[i] := y[i]*(1.0+spread*gasdev(idum));
      sig[i] := spread*y[i]
   END;
   mfit := 6;
   FOR i := 1 TO mfit DO lista[i] := i;
   alamda := -1;
   FOR i := 1 TO ma DO a[i] := gues[i];
   mrqmin(x,y,sig,npt,a,ma,lista,mfit,covar,alpha,chisq,alamda);
   k := 1;
   itst := 0;
   REPEAT
      writeln;
      writeln('Iteration #',k:2,'chi-squared:':17,chisq:10:4,
         'alamda:':10,alamda:9);
      writeln('a[1]':7,'a[2]':8,'a[3]':8,'a[4]':8,'a[5]':8,'a[6]':8);
      FOR i := 1 TO 6 DO write(a[i]:8:4);
      writeln;
      k := k+1;
      ochisq := chisq;
```

```
      mrqmin(x,y,sig,npt,a,ma,lista,mfit,covar,alpha,chisq,alamda);
      IF chisq > ochisq THEN
         itst := 0
      ELSE IF abs(ochisq-chisq) < 0.1 THEN
         itst := itst+1;
   UNTIL itst >=2;
   alamda := 0.0;
   mrqmin(x,y,sig,npt,a,ma,lista,mfit,covar,alpha,chisq,alamda);
   writeln('Uncertainties:');
   FOR i := 1 TO 6 DO write(sqrt(covar[i,i]):8:4);
   writeln
END.
```

fgauss is an example of the type of procedure that must be supplied to **mrqmin** in order to fit a user-defined function, in this case the sum of Gaussians. **fgauss** calculates both the function, and its derivative with respect to each adjustable parameter in a fairly compact fashion. The sample program **d14r10** calculates the same quantities in a more pedantic fashion, just to be sure we got everything right.

```
PROGRAM d14r10(input,output);
(* driver for routine FGAUSS *)
(*$I MODFILE.PAS *)
CONST
   npt = 3;
   nlin = 2;
   ma = 6;   (* ma=3*nlin *)
TYPE
   RealArrayMA = ARRAY [1..ma] OF real;
VAR
   e1,e2,f,x,y: real;
   i,j: integer;
   a,dyda,df: RealArrayMA;
(*$I FGAUSS.PAS *)
BEGIN
   a[1] := 3.0; a[2] := 0.2; a[3] := 0.5;
   a[4] := 1.0; a[5] := 0.7; a[6] := 0.3;
   writeln;
   writeln('x':6,'y':8,'dyda1':9,'dyda2':8,'dyda3':8,
      'dyda4':8,'dyda5':8,'dyda6':8);
   FOR i := 1 TO npt DO BEGIN
      x := 0.3*i;
      fgauss(x,a,y,dyda,ma);
      e1 := exp(-sqr((x-a[2])/a[3]));
      e2 := exp(-sqr((x-a[5])/a[6]));
      f := a[1]*e1+a[4]*e2;
      df[1] := e1;
      df[4] := e2;
      df[2] := a[1]*e1*2.0*(x-a[2])/(a[3]*a[3]);
      df[5] := a[4]*e2*2.0*(x-a[5])/(a[6]*a[6]);
      df[3] := a[1]*e1*2.0*sqr(x-a[2])/(a[3]*a[3]*a[3]);
      df[6] := a[4]*e2*2.0*sqr(x-a[5])/(a[6]*a[6]*a[6]);
      writeln('from FGAUSS');
      write(x:8:4,y:8:4);
      FOR j := 1 TO 6 DO write(dyda[j]:8:4);
      writeln;
      writeln('independent calc.');
      write(x:8:4,f:8:4);
```

```
         FOR j := 1 TO 6 DO write(df[j]:8:4);
         writeln;
         writeln
      END
END.
```

medfit is a procedure illustrating a more "robust" way of fitting. It performs a fit of data to a straight line, but instead of using the least-squares criterion for figuring the merit of a fit, it uses the least-absolute-deviation. For comparison, sample routine d14r11 fits lines to a noisy linear data set, using first the least-squares routine fit, and then the least-absolute-deviation routine medfit. You may be interested to see if you can figure out what mean value of absolute deviation you expect for data with gaussian noise of amplitude spread.

```
PROGRAM d14r11(input,output);
(* driver for routine MEDFIT *)
(*$I MODFILE.PAS *)
CONST
   ndata = 100;
   spread = 0.1;
TYPE
   RealArray55 = ARRAY [1..55] OF real;
   RealArrayNDATA = ARRAY [1..ndata] OF real;
   RealArrayNP = RealArrayNDATA;
VAR
   Ran3Inext,Ran3Inextp: integer;
   Ran3Ma: RealArray55;
   GasdevIset: integer;
   GasdevGset: real;
   a,abdev,b,chi2,q,siga,sigb: real;
   i,idum,mwt: integer;
   sig,x,y: RealArrayNDATA;
(*$I RAN3.PAS *)
(*$I GASDEV.PAS *)
(*$I GAMMLN.PAS *)
(*$I GSER.PAS *)
(*$I GCF.PAS *)
(*$I GAMMQ.PAS *)
(*$I SORT.PAS *)
(*$I FIT.PAS *)
(*$I MEDFIT.PAS *)
BEGIN
   GasdevIset := 0;
   idum := -1984;
   FOR i := 1 TO ndata DO BEGIN
      x[i] := 0.1*i;
      y[i] := -2.0*x[i]+1.0+spread*gasdev(idum);
      sig[i] := spread
   END;
   mwt := 1;
   writeln;
   fit(x,y,ndata,sig,mwt,a,b,siga,sigb,chi2,q);
   writeln('According to routine FIT the result is:');
   writeln('a = ':7,a:8:4,' uncertainty: ':13,siga:8:4);
   writeln('b = ':7,b:8:4,' uncertainty: ':13,sigb:8:4);
   writeln('chi-squared: ':16,chi2:8:4,' for ',ndata:4,' points');
```

```
        writeln('goodness-of-fit: ':20,q:8:4);
        writeln;
        writeln('According to routine MEDFIT the result is:');
        medfit(x,y,ndata,a,b,abdev);
        writeln('a = ':7,a:8:4);
        writeln('b = ':7,b:8:4);
        writeln(' ':3,'absolute deviation (per data point): ',abdev:8:4);
        writeln(' ':3,'note: gaussian spread is',spread:8:4,')')
END.
```

Chapter 15: Ordinary Differential Equations

Chapter 15 of Numerical Recipes deals with the integration of ordinary differential equations, restricting its attention specifically to initial-value problems. Three practical methods are introduced: 1) Runge-Kutta methods (rk4, rkdumb, rkqc, and odeint), 2) Richardson extrapolation and the Bulirsch-Stoer method (bsstep, mmid, rzextr, pzextr), 3) predictor-corrector methods. In general, for applications not demanding high precision, and where convenience is paramount, the fourth-order Runge-Kutta with adaptive step-size control is recommended. For higher precision applications, the Bulirsch-Stoer method dominates. The predictor-corrector methods are covered because of their history of widespread use, but are not regarded (by us) as having an important role today. (For a possible exception to this strong statement, see Numerical Recipes.)

$$\star \quad \star \quad \star \quad \star$$

Routine rk4 advances the solution vector y[n] of a set of ordinary differential equations over a single small interval h in x using the fourth-order Runge-Kutta method. The operation is shown by sample program d15r1 for an array of four variables y[1],...,y[4]. The first-order differential equations satisfied by these variables are specified by the accompanying routine derivs, and are simply the equations describing the first four Bessel functions $J_0(x),\ldots,J_3(x)$. The y's are initialized to the values of these functions at $x = 1.0$. Note that the values of dydx are also initialized at $x = 1.0$, because rk4 uses the values of dydx before its first call to derivs. The reason for this is discussed in the text. The sample program calls rk4 with h (the step-size) set to various values from 0.2 to 1.0, so that you can see how well rk4 can do even with quite sizeable steps.

```
PROGRAM d15r1(input,output);
(* driver for routine RK4 *)
(*$I MODFILE.PAS *)
CONST
   n = 4;
TYPE
   RealArrayNVAR = ARRAY [1..n] OF real;
VAR
   h,x: real;
   i,j: integer;
   y,dydx,yout: RealArrayNVAR;
(*$I BESSJ0.PAS *)
(*$I BESSJ1.PAS *)
(*$I BESSJ.PAS *)
PROCEDURE derivs(x: real; VAR y,dydx: RealArrayNVAR);
BEGIN
```

```
   dydx[1] := -y[2];
   dydx[2] := y[1]-(1.0/x)*y[2];
   dydx[3] := y[2]-(2.0/x)*y[3];
   dydx[4] := y[3]-(3.0/x)*y[4]
END;
(*$I RK4.PAS *)
BEGIN
   x := 1.0;
   y[1] := bessj0(x);
   y[2] := bessj1(x);
   y[3] := bessj(2,x);
   y[4] := bessj(3,x);
   dydx[1] := -y[2];
   dydx[2] := y[1]-y[2];
   dydx[3] := y[2]-2.0*y[3];
   dydx[4] := y[3]-3.0*y[4];
   writeln;
   writeln('Bessel function:':16,'j0':5,
         'j1':12,'j3':12,'j4':12);
   FOR i := 1 TO 5 DO BEGIN
      h := 0.2*i;
      rk4(y,dydx,n,x,h,yout);
      writeln;
      writeln('for a step size of:',h:6:2);
      write('rk4: ':11);
      FOR j := 1 TO 4 DO write(yout[j]:12:6);
      writeln;
      writeln('actual: ':11,bessj0(x+h):12:6,
            bessj1(x+h):12:6,bessj(2,x+h):12:6,bessj(3,x+h):12:6)
   END
END.
```

rkdumb is an extension of rk4 which allows you to integrate over larger intervals. It is "dumb" in the sense that it has no adaptive step-size determination, and no code to estimate errors. Sample program d15r2 works with the same functions and derivatives as the previous program, but integrates from x1=1.0 to x2=20.0, breaking the interval into nstep=150 equal steps. The variables vstart[1],..., vstart[4] which become the starting values of the y's, are initialized as before, but their derivatives this time are not initialized; rkdumb takes care of that. This time only the results for the fourth variable $J_3(x)$ are listed, and only every tenth value is given.

```
PROGRAM d15r2(input,output);
(* driver for routine RKDUMB *)
(*$I MODFILE.PAS *)
CONST
   nvar = 4;
   nstep = 150;
TYPE
   RealArrayNVAR = ARRAY [1..nvar] OF real;
   RealArray200 = ARRAY [1..200] OF real;
   RealArrayNVARby200 = ARRAY [1..nvar,1..200] OF real;
VAR
   i,j: integer;
   x1,x2: real;
   vstart: RealArrayNVAR;
   RkdumbX: RealArray200;
```

```
      RkdumbY: RealArrayNVARby200;
(*$I BESSJ0.PAS *)
(*$I BESSJ1.PAS *)
(*$I BESSJ.PAS *)
PROCEDURE derivs(x: real; VAR y,dydx: RealArrayNVAR);
BEGIN
   dydx[1] := -y[2];
   dydx[2] := y[1]-(1.0/x)*y[2];
   dydx[3] := y[2]-(2.0/x)*y[3];
   dydx[4] := y[3]-(3.0/x)*y[4]
END;
(*$I RK4.PAS *)
(*$I RKDUMB.PAS *)
BEGIN
   x1 := 1.0;
   vstart[1] := bessj0(x1);
   vstart[2] := bessj1(x1);
   vstart[3] := bessj(2,x1);
   vstart[4] := bessj(3,x1);
   x2 := 20.0;
   rkdumb(vstart,nvar,x1,x2,nstep);
   writeln('x':8,'integrated':17,'bessj3':10);
   FOR i := 1 TO nstep DIV 10 DO BEGIN
      j := 10*i;
      writeln(RkdumbX[j]:10:4,'  ',RkdumbY[4,j]:12:6,
         bessj(3,RkdumbX[j]):12:6)
   END
END.
```

rkqc performs a single step of fifth-order Runge-Kutta integration, this time with monitoring of local truncation error and corresponding step-size adjustment. Its sample program d15r3 is similar to that for routine rk4, using four Bessel functions as the example, and starting the integration at $x = 1.0$. However, on each pass a value is set for eps, the desired accuracy, and the trial value **htry** for the interval size is set to 0.1. For the first few passes, eps is not too demanding and **htry** may be perfectly adequate. As eps becomes smaller, the routine will be forced to diminish h and return smaller values of hdid and hnext. Our results (in single precision) are:

eps	htry	hdid	hnext
.3679E+00	.10	.100000	.400000
.1353E+00	.10	.100000	.354954
.4979E-01	.10	.100000	.287823
.1832E-01	.10	.100000	.233879
.6738E-02	.10	.100000	.190423
.2479E-02	.10	.100000	.155293
.9119E-03	.10	.100000	.126883
.3355E-03	.10	.100000	.103845
.1234E-03	.10	.073460	.066323
.4540E-04	.10	.034162	.031216
.1670E-04	.10	.028686	.025915
.6144E-05	.10	.011732	.010733
.2260E-05	.10	.010758	.009784
.8315E-06	.10	.004284	.003911
.3059E-06	.10	.004460	.004035

```
PROGRAM d15r3(input,output);
(* driver for routine RKQC *)
(*$I MODFILE.PAS *)
CONST
   n = 4;
TYPE
   RealArrayNVAR = ARRAY [1..n] OF real;
VAR
   eps,hdid,hnext,htry,x: real;
   i: integer;
   y,dydx,yscal: RealArrayNVAR;
(*$I BESSJ0.PAS *)
(*$I BESSJ1.PAS *)
(*$I BESSJ.PAS *)
PROCEDURE derivs(x: real; VAR y,dydx: RealArrayNVAR);
BEGIN
   dydx[1] := -y[2];
   dydx[2] := y[1]-(1.0/x)*y[2];
   dydx[3] := y[2]-(2.0/x)*y[3];
   dydx[4] := y[3]-(3.0/x)*y[4]
END;
(*$I RK4.PAS *)
(*$I RKQC.PAS *)
BEGIN
   x := 1.0;
   y[1] := bessj0(x);
   y[2] := bessj1(x);
   y[3] := bessj(2,x);
   y[4] := bessj(3,x);
   dydx[1] := -y[2];
   dydx[2] := y[1]-y[2];
   dydx[3] := y[2]-2.0*y[3];
   dydx[4] := y[3]-3.0*y[4];
   FOR i := 1 TO n DO yscal[i] := 1.0;
   htry := 0.1;
   writeln('eps':8,'htry':13,'hdid':12,'hnext':13);
   FOR i := 1 TO 15 DO BEGIN
      eps := exp(-i);
      rkqc(y,dydx,n,x,htry,eps,yscal,hdid,hnext);
      writeln(eps:13,htry:8:2,hdid:14:6,hnext:12:6)
   END
END.
```

The full driver routine for `rkqc`, which provides Runge-Kutta integration over large intervals with adaptive step-size control, is `odeint`. It plays the same role for `rkqc` that `rkdumb` plays for `rk4`. Integration is performed on four Bessel functions from `x1=1.0` to `x2=10.0`, with an accuracy `eps=1.0e-4`. Independent of the values of step-size actually used by `odeint`, intermediate values will be recorded only at intervals greater than `dxsav`. The sample program returns values of $J_3(x)$ for checking against actual values produced by `bessj`. It also records how many steps were successful, and how many were "bad". Bad steps are redone, and indicate no extra loss in accuracy. At the same time, they do represent a loss in efficiency, so that an excessive number of bad steps should initiate an investigation.

```pascal
PROGRAM d15r4(input,output);
(* driver for ODEINT *)
(*$I MODFILE.PAS *)
CONST
   n = 4;
TYPE
   RealArrayNVAR = ARRAY [1..n] OF real;
   RealArray200 = ARRAY [1..200] OF real;
   RealArrayNby200 = ARRAY [1..n,1..200] OF real;
VAR
   OdeintDxsav,eps,h1,hmin,x1,x2: real;
   i,OdeintKmax,OdeintKount,nbad,nok: integer;
   ystart: RealArrayNVAR;
   OdeintXp: RealArray200;
   OdeintYp: RealArrayNby200;
(*$I BESSJ0.PAS *)
(*$I BESSJ1.PAS *)
(*$I BESSJ.PAS *)
PROCEDURE derivs(x: real; VAR y,dydx: RealArrayNVAR);
(* Programs using routine DERIVS must define the type
TYPE
   RealArrayNVAR = ARRAY [1..4] OF real;
in the calling routine. *)
BEGIN
   dydx[1] := -y[2];
   dydx[2] := y[1]-(1.0/x)*y[2];
   dydx[3] := y[2]-(2.0/x)*y[3];
   dydx[4] := y[3]-(3.0/x)*y[4]
END;
(*$I RK4.PAS *)
(*$I RKQC.PAS *)
(*$I ODEINT.PAS *)
BEGIN
   x1 := 1.0;
   x2 := 10.0;
   ystart[1] := bessj0(x1);
   ystart[2] := bessj1(x1);
   ystart[3] := bessj(2,x1);
   ystart[4] := bessj(3,x1);
   eps := 1.0e-4;
   h1 := 0.1;
   hmin := 0.0;
   OdeintKmax := 100;
   OdeintDxsav := (x2-x1)/20.0;
   odeint(ystart,n,x1,x2,eps,h1,hmin,nok,nbad);
   writeln;
   writeln('successful steps:',' ':13,nok:3);
   writeln('bad steps:',' ':20,nbad:3);
   writeln('stored intermediate values:',' ',OdeintKount:3);
   writeln;
   writeln('x':8,'integral':18,'bessj(3,x)':15);
   FOR i := 1 TO OdeintKount DO
      writeln(OdeintXp[i]:10:4,OdeintYp[4,i]:16:6,bessj(3,OdeintXp[i]):14:6)
END.
```

The modified midpoint routine mmid is presented in *Numerical Recipes* primarily as a component of the more powerful Bulirsch-Stoer routine. It integrates variables over an interval htot through a sequence of much smaller steps. Sample routine d15r5 takes the number of subintervals i to be $5, 10, 15, \ldots, 50$ so that we can witness any improvements in accuracy that may occur. The values of the four Bessel functions are compared with the results of the integrations.

```pascal
PROGRAM d15r5(input,output);
(* driver for routine MMID *)
(*$I MODFILE.PAS *)
CONST
    nvar = 4;
    x1 = 1.0;
    htot = 0.5;
TYPE
    RealArrayNVAR = ARRAY [1..nvar] OF real;
VAR
    b1,b2,b3,b4,xf: real;
    i,ii: integer;
    y,yout,dydx: RealArrayNVAR;
(*$I BESSJ0.PAS *)
(*$I BESSJ1.PAS *)
(*$I BESSJ.PAS *)
PROCEDURE derivs(x: real; VAR y,dydx: RealArrayNVAR);
(* Programs using DERIVS must define the type
TYPE
    RealArrayNVAR = ARRAY [1..4] OF real;
in the calling routine. *)
BEGIN
    dydx[1] := -y[2];
    dydx[2] := y[1]-(1.0/x)*y[2];
    dydx[3] := y[2]-(2.0/x)*y[3];
    dydx[4] := y[3]-(3.0/x)*y[4]
END;
(*$I MMID.PAS *)
BEGIN
    y[1] := bessj0(x1);
    y[2] := bessj1(x1);
    y[3] := bessj(2,x1);
    y[4] := bessj(3,x1);
    dydx[1] := -y[2];
    dydx[2] := y[1]-y[2];
    dydx[3] := y[2]-2.0*y[3];
    dydx[4] := y[3]-3.0*y[4];
    xf := x1+htot;
    b1 := bessj0(xf);
    b2 := bessj1(xf);
    b3 := bessj(2,xf);
    b4 := bessj(3,xf);
    writeln('First four Bessel functions:');
    FOR ii := 1 TO 10 DO BEGIN
        i := 5*ii;
        mmid(y,dydx,nvar,x1,htot,i,yout);
        writeln;
        writeln('x := ',x1:5:2,
            ' to ',x1+htot:5:2,' in ',i:2,' steps');
        writeln('integration':14,'bessj':9);
        writeln(yout[1]:12:6,b1:12:6);
        writeln(yout[2]:12:6,b2:12:6);
```

```
        writeln(yout[3]:12:6,b3:12:6);
        writeln(yout[4]:12:6,b4:12:6);
        writeln('press return to continue...');
        readln
    END
END.
```

The Bulirsch-Stoer method, illustrated by routine **bsstep**, is the integrator of choice for higher accuracy calculations on smooth functions. An interval h is broken into finer and finer steps, and the results of integration are extrapolated to zero step-size. The extrapolation is via a rational function with **rzextr**. **bsstep** monitors local truncation error and adjusts the step-size appropriately, to keep errors below **eps**. From an external point of view, **bsstep** operates exactly as does **rkqc**: it has the same arguments and in the same order. Consequently it can be used in place of **rkqc** in routine **odeint**, allowing more efficient integration over large regions of x. For this reason, the sample program d15r6 is of the same form used to demonstrate **rkqc**.

```
PROGRAM d15r6(input,output);
(* driver for routine BSSTEP *)
(*$I MODFILE.PAS *)
CONST
    n = 4;
    RzextrImax = 11;
    RzextrNmax = 10;
    RzextrNcol = 7;
TYPE
    RealArrayNVAR = ARRAY [1..n] OF real;
VAR
    RzextrX: ARRAY [1..RzextrImax] OF real;
    RzextrD: ARRAY [1..RzextrNmax,1..RzextrNcol] OF real;
    eps,hdid,hnext,htry,x: real;
    i: integer;
    y,dydx,yscal: RealArrayNVAR;
PROCEDURE derivs(x: real; y: RealArrayNVAR; VAR dydx: RealArrayNVAR);
(* Programs using DERIVS must define the type
TYPE
    RealArrayNVAR = ARRAY [1..4] OF real;
in the main routine. *)
BEGIN
    dydx[1] := -y[2];
    dydx[2] := y[1]-(1.0/x)*y[2];
    dydx[3] := y[2]-(2.0/x)*y[3];
    dydx[4] := y[3]-(3.0/x)*y[4]
END;
(*$I BESSJ0.PAS *)
(*$I BESSJ1.PAS *)
(*$I BESSJ.PAS *)
(*$I MMID.PAS *)
(*$I RZEXTR.PAS *)
(*$I BSSTEP.PAS *)
BEGIN
    x := 1.0;
    y[1] := bessj0(x);
    y[2] := bessj1(x);
    y[3] := bessj(2,x);
    y[4] := bessj(3,x);
```

```
    dydx[1] := -y[2];
    dydx[2] := y[1]-y[2];
    dydx[3] := y[2]-2.0*y[3];
    dydx[4] := y[3]-3.0*y[4];
    FOR i := 1 TO n DO yscal[i] := 1.0;
    htry := 1.0;
    writeln;
    writeln('eps':10,'htry':12,'hdid':12,'hnext':12);
    FOR i := 1 TO 15 DO BEGIN
        eps := exp(-i);
        bsstep(y,dydx,n,x,htry,eps,yscal,hdid,hnext);
        writeln('    ',eps:11,htry:8:2,hdid:14:6,hnext:12:6)
    END
END.
```

rzextr performs a diagonal rational function extrapolation for bsstep. It takes
a sequence of interval lengths and corresponding integrated values, and extrapolates
to the value the integral would have if the interval length were zero. Sample routine
d15r7 works with a known function

$$F_n = \frac{1 - x + x^3}{(x+1)^n} \qquad n = 1,..,4$$

We extrapolate the vector yest $= (F_1, F_2, F_3, F_4)$ given a sequence of ten values (only
the last nuse=5 of which are used). The ten values are labelled iest=1,...,10 and are
evaluated at xest=1.0/iest. A call to rzextr produces extrapolated values yz, and
estimated errors dy, and compares to the true values $(1.0, 1.0, 1.0, 1.0)$ at xest=0.0.

```
PROGRAM d15r7(input,output);
(* driver for routine RZEXTR *)
(*$I MODFILE.PAS *)
CONST
    RzextrImax = 11;
    RzextrNmax = 10;
    RzextrNcol = 7;
    nvar = 4;
    nuse = 5;
TYPE
    RealArrayNVAR = ARRAY [1..nvar] OF real;
VAR
    RzextrX: ARRAY [1..RzextrImax] OF real;
    RzextrD: ARRAY [1..RzextrNmax,1..RzextrNcol] OF real;
    dum,xest: real;
    i,iest,j: integer;
    dy,yest,yz: RealArrayNVAR;
(*$I RZEXTR.PAS *)
BEGIN
(* feed values from a rational function *)
(* fn(x) := (1-x+x**3)/(x+1)**n *)
    FOR i := 1 TO 10 DO BEGIN
        iest := i;
        xest := 1.0/i;
        dum := 1.0-xest+xest*xest*xest;
        FOR j := 1 TO nvar DO BEGIN
            dum := dum/(xest+1.0);
            yest[j] := dum
        END;
```

```
        rzextr(iest,xest,yest,yz,dy,nvar,nuse);
        writeln;
        writeln('iest  :=  ',i:2,'   xest =',xest:8:4);
        write('Extrap. function: ');
        FOR j := 1 TO nvar DO write(yz[j]:12:6);
        writeln;
        write('Estimated error:  ');
        FOR j := 1 TO nvar DO write(dy[j]:12:6);
        writeln
     END;
     writeln;
     writeln('Actual values:     ',1.0:12:6,1.0:12:6,1.0:12:6,1.0:12:6)
  END.
```

pzextr is a less powerful standby for rzextr, to be used primarily when some problem crops up with the extrapolation. It performs a polynomial, rather than a rational function, extrapolation. The sample program d15r8 is identical to that for rzextr.

```
PROGRAM d15r8(input,output);
(* driver for routine PZEXTR *)
(*$I MODFILE.PAS *)
CONST
   PzextrImax = 11;
   PzextrNmax = 10;
   PzextrNcol = 7;
   nvar = 4;
   nuse = 5;
TYPE
   RealArrayNVAR = ARRAY [1..nvar] OF real;
VAR
   PzextrX: ARRAY [1..PzextrImax] OF real;
   PzextrQcol: ARRAY [1..PzextrNmax,1..PzextrNcol] OF real;
   dum,xest: real;
   i,iest,j: integer;
   dy,yest,yz: RealArrayNVAR;
(*$I PZEXTR.PAS *)
BEGIN
(* feed values from a rational function *)
(* fn(x) := (1-x+x**3)/(x+1)**n *)
   FOR i := 1 TO 10 DO BEGIN
      iest := i;
      xest := 1.0/i;
      dum := 1.0-xest+xest*xest*xest;
      FOR j := 1 TO nvar DO BEGIN
         dum := dum/(xest+1.0);
         yest[j] := dum
      END;
      pzextr(iest,xest,yest,yz,dy,nvar,nuse);
      writeln;
      writeln('i  :=  ',i:2);
      write('Extrap. function:');
      FOR j := 1 TO nvar DO write(yz[j]:12:6);
      writeln;
      write('Estimated error: ');
      FOR j := 1 TO nvar DO write(dy[j]:12:6);
      writeln
```

```
        END;
        writeln;
        writeln('actual values:    ',1.0:12:6,1.0:12:6,1.0:12:6,1.0:12:6)
END.
```

Chapter 16: Two-Point Boundary Value Problems

Two-point boundary value problems, and their iterative solution, is the substance of Chapter 16 of Numerical Recipes. The first step is to cast the problem as a set of N coupled first-order ordinary differential equations, satisfying n_1 conditions at one boundary point, and $n_2 = N - n_1$ conditions at the other boundary point. We apply two general methods to the solutions. First are the shooting methods, typified by procedures shoot and shootf, which enforce the n_1 conditions at one boundary and set n_2 conditions freely. Then they integrate across the interval to find discrepancies with the n_2 conditions at the other end. The Newton-Raphson method is used to reduce these discrepancies by adjusting the variable parameters.

The other approach is the relaxation method in which the differential equations are replaced by finite difference equations on a grid that covers the range of interest. Routine solvde demonstrates this method, and is demonstrated "in action" by program sfroid, which uses it to compute eigenvalues of spheroidal harmonics. Since the program sfroid in Numerical Recipes is already self-contained, we need concern ourselves here only with shooting routines. For the purpose of comparison, we apply these routines to the same problem attacked with sfroid.

$$\star \quad \star \quad \star \quad \star$$

Procedure shoot works as described above. Demonstration program d16r1 uses it to find eigenvalues of both prolate and oblate spheroidal harmonics. The oblate and prolate cases are handled simultaneously, although they actually involve two independent sets of three coupled first-order differential equations, one set with c^2 positive and the other with c^2 negative. The complete set of differential equations is

$$\frac{dy_1}{dx} = y_2$$

$$\frac{dy_2}{dx} = \frac{2x(m+1)y_2 - (y_3 - c^2x^2)y_1}{(1 - x^2)}$$

$$\frac{dy_3}{dx} = 0$$

$$\frac{dy_4}{dx} = y_5$$

$$\frac{dy_5}{dx} = \frac{2x(m+1)y_5 - (y_6 + c^2x^2)y_4}{(1 - x^2)}$$

$$\frac{dy_6}{dx} = 0$$

These are specified in procedure derivs which is called, in turn, by odeint in shoot. The first three equations correspond to prolate harmonics and the second three to oblate harmonics. Comparing either set of three to equation (16.4.4) in *Numerical Recipes*, you may quickly verify that y_1 and y_4 correspond to the two spheroidal harmonic solutions, y_3 and y_6 correspond to the sought-after eigenvalues (whose derivative with respect to x is of course 0), and y_2 and y_5 are intermediate variables created to change the second-order equations to coupled first-order equations.

Two other procedures are used by shoot. Procedure load sets the values of all the variables $y_1, \ldots, y_6$ at the first boundary, and score calculates a discrepancy vector F (which will be zero when a successful solution has been reached) at the second boundary. Each of these procedures has some interesting aspects. In load, y_3 and y_6 are initialized to v[1] and v[2], values calculated in the sample program to give rough estimates of the size of the proper result. We arrived at these estimates just by looking through some tables of values. Also notice that, for example, y_1 is set to factr + $y_2$dx. This is the same as saying that $y_1 = $ factr + $(dy_1/dx)\Delta x$. The quantity factr comes from equation (16.4.20) in *Numerical Recipes*, and the term with dx comes from the fact that we placed the lower boundary x1 at $-1.0 + $ dx (where dx=1.0e-4) rather than at -1.0. This is because dy_2/dx and dy_5/dx cannot be evaluated exactly at $x = -1.0$. The procedure score follows from equation (16.4.18) in *Numerical Recipes*. For example, if $N - M$ is odd, $y_1 = 0$ at $x = 0$, but if $N - M$ is even, then $y_2 = dy_1/dx = 0$.

That more or less explains things. Now, given M, N, and c^2, sample program d16r1 sets up estimates v[1] and v[2] and iterates the routine shoot until changes in the v are less than some preset fraction eps of their size. Some values of the eigenvalues of the spheroidal harmonics are given in section 16.4 of *Numerical Recipes* if you want to check the results.

```
PROGRAM d16r1(input,output);
(* driver for routine SHOOT *)
(* Solves for eigenvalues of spheroidal harmonics. Both
prolate and oblate case are handled simultaneously, leading
to six first-order equations. Unknown to shoot, it is
actually two independent sets of three coupled equations,
one set with c^2 positive and the other with c^2 negative.  *)
(*$I MODFILE.PAS *)
CONST
   nvar = 6;
   n2 = 2;
   delta = 1.0e-3;
   eps = 1.0e-6;
   dx = 1.0e-4;
TYPE
   RealArrayN2 = ARRAY [1..n2] OF real;
   RealArrayNVAR = ARRAY [1..nvar] OF real;
   RealArrayN2byN2 = ARRAY [1..n2,1..n2] OF real;
   RealArrayNP = RealArrayN2;
   IntegerArrayNP = ARRAY [1..n2] OF integer;
   RealArrayNPbyNP = RealArrayN2byN2;
   RealArray200 = ARRAY [1..200] OF real;
   RealArrayNby200 = ARRAY [1..nvar,1..200] OF real;
VAR
   ShootC2,ShootFactr,h1,hmin,q1,x1,x2: real;
```

```
   i,ShootM,ShootN: integer;
   delv,v: RealArrayN2;
   dv,f: RealArrayNP;
   OdeintKmax,OdeintKount: integer;
   OdeintDxsav: real;
   OdeintXp: RealArray200;
   OdeintYp: RealArrayNby200;
PROCEDURE load(x1: real; VAR v: RealArrayN2; VAR y: RealArrayNVAR);
(* Programs using routine LOAD must declare the global variables
VAR
   ShootC2,ShootFactr: real;
   ShootM: integer;
in the main routine. *)
BEGIN
   y[3] := v[1];
   y[2] := -(y[3]-ShootC2)*ShootFactr/2.0/(ShootM+1.0);
   y[1] := ShootFactr+y[2]*dx;
   y[6] := v[2];
   y[5] := -(y[6]+ShootC2)*ShootFactr/2.0/(ShootM+1.0);
   y[4] := ShootFactr+y[5]*dx
END;
PROCEDURE score(x2: real; VAR y: RealArrayNVAR; VAR f: RealArrayN2);
(* Programs using routine SCORE must declare the global variables
VAR
   ShootM,ShootN: integer;
in the main routine. *)
BEGIN
   IF odd(ShootN-ShootM) THEN BEGIN
       f[1] := y[1];
       f[2] := y[4]
   END ELSE BEGIN
       f[1] := y[2];
       f[2] := y[5]
   END
END;
PROCEDURE derivs(x: real; VAR y,dydx: RealArrayNVAR);
(* Programs using routine DERIVS must declare the global variables
VAR
   ShootC2: real;
   ShootM: integer;
in the main routine. *)
BEGIN
   dydx[1] := y[2];
   dydx[3] := 0.0;
   dydx[2] := (2.0*x*(ShootM+1.0)*y[2]-(y[3]-ShootC2*x*x)*y[1])/(1.0-x*x);
   dydx[4] := y[5];
   dydx[6] := 0.0;
   dydx[5] := (2.0*x*(ShootM+1.0)*y[5]-(y[6]+ShootC2*x*x)*y[4])/(1.0-x*x)
END;
(*$I LUBKSB.PAS *)
(*$I LUDCMP.PAS *)
(*$I RK4.PAS *)
(*$I RKQC.PAS *)
(*$I ODEINT.PAS *)
(*$I SHOOT.PAS *)
BEGIN
   REPEAT
```

```
    write('Input M,N,C-Squared:  ');
    readln(ShootM,ShootN,ShootC2);
UNTIL (ShootN >= ShootM) AND (ShootM >= 0);
ShootFactr := 1.0;
IF ShootM <> 0 THEN BEGIN
    q1 := ShootN;
    FOR i := 1 TO ShootM DO BEGIN
        ShootFactr := -0.5*ShootFactr*(ShootN+i)*(q1/i);
        q1 := q1-1.0
    END
END;
v[1]  := ShootN*(ShootN+1)-ShootM*(ShootM+1)+ShootC2/2.0;
v[2]  := ShootN*(ShootN+1)-ShootM*(ShootM+1)-ShootC2/2.0;
delv[1] := delta*v[1];
delv[2] := delv[1];
h1  := 0.1;
hmin := 0.0;
x1  := -1.0+dx;
x2  := 0.0;
writeln;
writeln('Prolate':17,'Oblate':23);
writeln('Mu(m,n)':11,'Error Est.':14,'Mu(m,n)':10,'Error Est.':14);
REPEAT
    shoot(nvar,v,delv,n2,x1,x2,eps,h1,hmin,f,dv);
    writeln(v[1]:12:6,dv[1]:12:6,v[2]:12:6,dv[2]:12:6);
UNTIL (abs(dv[1]) <= abs(eps*v[1])) AND (abs(dv[2]) <= abs(eps*v[2]))
END.
```

Another shooting method is shooting to a fitting point. More explicitly, we set values at two boundaries, from both of which we integrate toward an intermediate point. For the spheroidal harmonics, we take the endpoints, in sample program d16r2, to be $-1.0+dx$ and $1.0-dx$, and the intermediate point to be $x = 0.0$. For clarity, we considered only prolate spheroids. The calculation is similar to that in the previous sample program, except for these details:

1. There are only three first-order differential equations in **derivs** because of the restriction to prolate spheroids. (Note: the oblate case requires only that we input c^2 as a negative number.)

2. There are two load routines, **load1** and **load2**, which set values at the two boundaries. At the first boundary y[3] is initialized to v1[1], which is initially set to our crude guess of the magnitude of the eigenvalue. y[1], the spheroidal harmonic value itself, is set to **factr** + $(dy_1/dx)\Delta x$, and y[2] is also set as before. At boundary two, y[3] and y[1] are given guessed values for the eigenvalue and for $y(1 - \Delta x)$ respectively. We treat the guessed eigenvalue at boundary two as independent of that at boundary one, although they ought certainly to converge to the same value. To verify this point, we make the initial guess that the values differ by 1.0 (i.e. v2[2]=v1[1]+1.0).

Sample program d16r2 otherwise proceeds much as d16r1 did, however with **score** kept at $x = 0.0$ where the solutions must match up. The procedure **score** has been set to a dummy operation equating F_i to y_i so that the condition of success is that the y_i all match at $x = 0$. This is discussed more fully in *Numerical Recipes*. Check the eigenvalue results against the previous routine.

```
PROGRAM d16r2(input,output);
(* driver for routine SHOOTF *)
(*$I MODFILE.PAS *)
LABEL 1;
CONST
   nvar = 3;
   n1 = 2;
   n2 = 1;
   delta = 1.0e-3;
   eps = 1.0e-6;
   dx = 1.0e-4;
TYPE
   RealArrayN1 = ARRAY [1..n1] OF real;
   RealArrayN2 = ARRAY [1..n2] OF real;
   RealArrayNVAR = ARRAY [1..nvar] OF real;
   RealArrayNP = RealArrayNVAR;
   RealArrayNVARbyNVAR = ARRAY [1..nvar,1..nvar] OF real;
   RealArrayNPbyNP = RealArrayNVARbyNVAR;
   IntegerArrayNP = ARRAY [1..nvar] OF integer;
   RealArray200 = ARRAY [1..200] OF real;
   RealArrayNby200 = ARRAY [1..nvar,1..200] OF real;
VAR
   ShootfC2,ShootfFactr,h1,hmin: real;
   q1,x1,x2,xf: real;
   i,ShootfM,ShootfN: integer;
   v1,delv1,dv1: RealArrayN2;
   v2,delv2,dv2: RealArrayN1;
   f: RealArrayNVAR;
   OdeintKmax,OdeintKount: integer;
   OdeintDxsav: real;
   OdeintXp: RealArray200;
   OdeintYp: RealArrayNby200;
PROCEDURE load1(x1: real; VAR v1: RealArrayN2; VAR y: RealArrayNVAR);
(* Programs using routine LOAD1 must declare the variables
VAR
   ShootfC2,ShootfFactr: real;
   ShootfM: integer;
in the main routine. *)
BEGIN
   y[3] := v1[1];
   y[2] := -(y[3]-ShootfC2)*ShootfFactr/2.0/(ShootfM+1.0);
   y[1] := ShootfFactr+y[2]*dx
END;
PROCEDURE load2(x2: real; VAR v2: RealArrayN1; VAR y: RealArrayNVAR);
(* Programs using routine LOAD2 must declare the variables
   ShootfC2: real;
   ShootfM: integer;
in the main routine. *)
BEGIN
   y[3] := v2[2];
   y[1] := v2[1];
   y[2] := (y[3]-ShootfC2)*y[1]/2.0/(ShootfM+1.0)
END;
PROCEDURE score(xf: real; VAR y,f: RealArrayNVAR);
VAR
   i: integer;
BEGIN
```

```
    FOR i := 1 TO 3 DO f[i] := y[i]
END;
PROCEDURE derivs(x: real; VAR y,dydx: RealArrayNVAR);
(* Programs using routine DERIVS must declare the variables
    ShootfC2: real;
    ShootfM: integer;
in the main routine. *)
BEGIN
    dydx[1] := y[2];
    dydx[3] := 0.0;
    dydx[2] := (2.0*x*(ShootfM+1.0)*y[2]-(y[3]-ShootfC2*x*x)*y[1])/(1.0-x*x)
END;
(*$I LUBKSB.PAS *)
(*$I LUDCMP.PAS *)
(*$I RK4.PAS *)
(*$I RKQC.PAS *)
(*$I ODEINT.PAS *)
(*$I SHOOTF.PAS *)
BEGIN
1: write('Input M,N,C-SQUARED: ');
    readln(ShootfM,ShootfN,ShootfC2);
    IF (ShootfN < ShootfM) OR (ShootfM < 0) THEN BEGIN
        writeln('Improper arguments');
        GOTO 1
    END;
    ShootfFactr := 1.0;
    IF ShootfM <> 0 THEN BEGIN
        q1 := ShootfN;
        FOR i := 1 TO ShootfM DO BEGIN
            ShootfFactr := -0.5*ShootfFactr*(ShootfN+i)*(q1/i);
            q1 := q1-1.0
        END
    END;
    v1[1] := ShootfN*(ShootfN+1)-ShootfM*(ShootfM+1)+ShootfC2/2.0;
    IF odd(ShootfN-ShootfM) THEN
        v2[1] := -ShootfFactr
    ELSE
        v2[1] := ShootfFactr;
    v2[2] := v1[1]+1.0;
    delv1[1] := delta*v1[1];
    delv2[1] := delta*ShootfFactr;
    delv2[2] := delv1[1];
    h1 := 0.1;
    hmin := 0.0;
    x1 := -1.0+dx;
    x2 := 1.0-dx;
    xf := 0.0;
    writeln;
    writeln('mu(-1)':26,'y(1-dx)':20,'mu(+1)':19);
    REPEAT
        shootf(nvar,v1,v2,delv1,delv2,n1,n2,x1,x2,
            xf,eps,h1,hmin,f,dv1,dv2);
        writeln;
        writeln('v ':6,v1[1]:20:6,v2[1]:20:6,v2[2]:20:6);
        writeln('dv':6,dv1[1]:20:6,dv2[1]:20:6,dv2[2]:20:6);
    UNTIL abs(dv1[1]) <= abs(eps*v1[1])
END.
```

Chapter 17: Partial Differential Equations

Several methods for solving partial differential equations by numerical means are treated in Chapter 17 of Numerical Recipes. All are finite differencing methods, including forward time centered space differencing, the Lax method, staggered leapfrog differencing, the two-step Lax-Wendroff scheme, the Crank-Nicholson method, Fourier analysis and cyclic reduction (FACR), Jacobi's method, the Gauss-Seidel method, simultaneous over-relaxation (SOR) with and without Chebyshev acceleration, and operator splitting methods as exemplified by the alternating direction implicit (ADI) method. There are so many methods, in fact, that we have not provided each topic with a procedure of its own. In many cases the nature of such procedures follows naturally from the description. In other cases, you will have to consult other references. The procedures that do appear in the chapter, sor and adi, show two of the more useful and efficient methods for elliptic equations in application.

$$\star \quad \star \quad \star \quad \star$$

Procedure **sor** incorporates simultaneous over-relaxation with Chebyshev acceleration to solve an elliptic partial differential equation. As input it accepts six arrays of coefficients, an estimate of the spectral radius of Jacobi iteration, and a trial solution which is often just set to zero over the solution grid. In program d17r1 the method is applied to the model problem

$$\frac{\partial^2 u}{\partial x^2} + \frac{\partial^2 u}{\partial y^2} = \rho$$

which is treated as the relaxation problem

$$\frac{\partial u}{\partial t} = \frac{\partial^2 u}{\partial x^2} + \frac{\partial^2 u}{\partial y^2} - \rho$$

Using FTCS differencing, this becomes

$$u_{j+1,l}^n + u_{j-1,l}^n + u_{j,l+1}^n + u_{j,l-1}^n - 4u_{j,l}^{n+1} = \rho_{jl}\Delta^2$$

(The notation is explained in Chapter 17 of *Numerical Recipes*.) This is a simple form of the general difference equation to which **sor** may be applied, with

$$A_{jl} = B_{jl} = C_{jl} = D_{jl} = 1.0 \text{ and } E_{jl} = -4.0$$

for all j and l. The starting guess for u is $u_{jl} = 0.0$ for all j, l. For a source function $F_{j,l}$ we took $F_{j,l} = 0.0$ except directly in the center of the grid where $F(\texttt{midl}, \texttt{midl}) =$

2.0. The value of ρ_{Jacobi}, which is called rjac, is taken from equation (17.5.24) of *Numerical Recipes*,

$$\rho_{Jacobi} = \frac{\cos\dfrac{\pi}{J} + \left(\dfrac{\Delta x}{\Delta y}\right)^2 \cos\dfrac{\pi}{L}}{1 + \left(\dfrac{\Delta x}{\Delta y}\right)^2}$$

In this case, j=l=jmax and $\Delta x = \Delta y$ so rjac $= \cos(\pi/\text{jmax})$. A call to sor leads to the solution shown below. As a test that this is indeed a solution to the finite difference equation, the program plugs the result back into that equation, calculating

$$F_{j,l} = u_{j+1,l}^n + u_{j-1,l}^n + u_{j,l+1}^n + u_{j,l-1}^n - 4u_{j,l}^{n+1}$$

The test is whether $F_{j,l}$ is almost everywhere zero, but equal to 2.0 at the very centerpoint of the grid.

```
SOR solution grid:
   .00    .00    .00    .00    .00    .00    .00    .00    .00    .00    .00
   .00  -.02  -.04  -.06  -.08  -.09  -.08  -.06  -.04  -.02    .00
   .00  -.04  -.09  -.13  -.17  -.19  -.17  -.13  -.09  -.04    .00
   .00  -.06  -.13  -.20  -.28  -.32  -.28  -.20  -.13  -.06    .00
   .00  -.08  -.17  -.28  -.41  -.55  -.41  -.28  -.17  -.08    .00
   .00  -.09  -.19  -.32  -.55 -1.05  -.55  -.32  -.19  -.09    .00
   .00  -.08  -.17  -.28  -.41  -.55  -.41  -.28  -.17  -.08    .00
   .00  -.06  -.13  -.20  -.28  -.32  -.28  -.20  -.13  -.06    .00
   .00  -.04  -.09  -.13  -.17  -.19  -.17  -.13  -.09  -.04    .00
   .00  -.02  -.04  -.06  -.08  -.09  -.08  -.06  -.04  -.02    .00
   .00    .00    .00    .00    .00    .00    .00    .00    .00    .00    .00
```

```pascal
PROGRAM d17r1(input,output);
(* driver for routine SOR *)
(*$I MODFILE.PAS *)
CONST
   jmax = 11;
   pi = 3.1415926;
TYPE
   DoubleArrayJMAXbyJMAX = ARRAY [1..jmax,1..jmax] OF double;
VAR
   i,j,midl: integer;
   rjac: double;
   a,b,c,d,e,f,u: DoubleArrayJMAXbyJMAX;
(*$I SOR.PAS *)
BEGIN
   FOR i := 1 TO jmax DO
      FOR j := 1 TO jmax DO BEGIN
         a[i,j] := 1.0;
         b[i,j] := 1.0;
         c[i,j] := 1.0;
         d[i,j] := 1.0;
         e[i,j] := -4.0;
         f[i,j] := 0.0;
         u[i,j] := 0.0
      END;
   midl := (jmax DIV 2)+1;
   f[midl,midl] := 2.0;
   rjac := cos(pi/jmax);
```

```
sor(a,b,c,d,e,f,u,jmax,rjac);
writeln('SOR Solution:');
FOR i := 1 TO jmax DO BEGIN
   FOR j := 1 TO jmax DO write(u[i,j]:7:2);
   writeln
END;
writeln;
writeln('Test that solution satisfies difference eqns:');
FOR i := 2 TO jmax-1 DO BEGIN
   FOR j := 2 TO jmax-1 DO
      f[i,j] := u[i+1,j]+u[i-1,j]+u[i,j+1]
         +u[i,j-1]-4.0*u[i,j];
   write(' ':7);
   FOR j := 2 TO jmax-1 DO write(f[i,j]:7:2);
   writeln
END
END.
```

Routine `adi` uses the alternating direction implicit method for solving partial differential equations. This method can be considerably more efficient than the `sor` calculation, and is preferred among relaxation methods when the shape of the grid and the boundary conditions allow its use. It is admittedly slightly more difficult to program, and sometimes does not converge, but it is the recommended "first-try" algorithm. Sample program d17r2 uses the same model problem outlined above. When it is subjected to operator splitting and put in the form of equations (17.6.22) of *Numerical Recipes*, the coefficient arrays become

$$A_{jl} = C_{jl} = D_{jl} = F_{jl} = -1.0$$

$$B_{jl} = E_{jl} = 2.0$$

Again the trial solution is set to zero everywhere, and the source term is zeroed except at the centerpoint of the grid. As given in the text (equation 17.6.20) bounds on the eigenvalues are

$$\text{alpha} = 2\left[1 - \cos\left(\frac{\pi}{\text{jmax}}\right)\right]$$

$$\text{beta} = 2\left[1 - \cos\frac{(\text{jmax}-1)\pi}{\text{jmax}}\right]$$

where `jmax` × `jmax` is the dimension of the grid. The number of iterations 2^k is minimized by choosing it to be about $\ln(4\text{jmax}/\pi)$. As in routine `sor`, the solution for u is printed out and may be compared with the copy listed before program d17r1. Also, this solution is substituted into the difference equation and should give a zero result everywhere except at the centerpoint of the grid, where its value is 2.0. Notice that `adi` makes calls to `tridag` and requires a double precision version of that routine, if available.

```
PROGRAM d17r2(input,output);
(* driver for routine ADI *)
(*$I MODFILE.PAS *)
CONST
   jmax = 11;
   pi = 3.1415926;
TYPE
```

```
      DoubleArrayJMAXbyJMAX = ARRAY [1..jmax,1..jmax] OF double;
VAR
   alim,alpha,beta,eps: double;
   i,j,k,mid,twotok: integer;
   a,b,c,d,e,f,g,u: DoubleArrayJMAXbyJMAX;
(*$I ADI.PAS *)
BEGIN
   FOR i := 1 TO jmax DO BEGIN
      FOR j := 1 TO jmax DO BEGIN
         a[i,j] := -1.0;
         b[i,j] := 2.0;
         c[i,j] := -1.0;
         d[i,j] := -1.0;
         e[i,j] := 2.0;
         f[i,j] := -1.0;
         g[i,j] := 0.0;
         u[i,j] := 0.0
      END
   END;
   mid := (jmax DIV 2)+1;
   g[mid,mid] := 2.0;
   alpha := 2.0*(1.0-cos(pi/jmax));
   beta := 2.0*(1.0-cos((jmax-1)*pi/jmax));
   alim := ln(4.0*jmax/pi);
   k := 0;
   twotok := 1;
   REPEAT
      k := k+1;
      twotok := 2*twotok;
   UNTIL twotok >= alim;
   eps := 1.0e-4;
   adi(a,b,c,d,e,f,g,u,jmax,k,alpha,beta,eps);
   writeln('ADI Solution:');
   FOR i := 1 TO jmax DO BEGIN
      FOR j := 1 TO jmax DO write(u[i,j]:7:2);
      writeln
   END;
   writeln;
   writeln('Test that solution satisfies difference eqns:');
   FOR i := 2 TO jmax-1 DO BEGIN
      FOR j := 2 TO jmax-1 DO
         g[i,j] := -4.0*u[i,j]+u[i+1,j]
            +u[i-1,j]+u[i,j-1]+u[i,j+1];
      write(' ':7);
      FOR j := 2 TO jmax-1 DO write(g[i,j]:7:2);
      writeln
   END
END.
```

Index of Demonstrated Procedures

Following is a brief explanation of each *Numerical Recipes* product, plus two order forms (one for United States and Canadian residents, one for all other), which may be used to order these items directly from the publisher if you cannot obtain them from your local bookstore.

Numerical Recipes (FORTRAN Version), Numerical Recipes in Pascal, and *Numerical Recipes in C* represent the main text and reference component of the *Numerical Recipes* package. Each book contains over 200 programs, in the language of the reader's choice, and constitutes a complete subroutine library for scientific computation. All versions contain equivalent tutorial, mathematical, and practical discussions.

There are three example books containing FORTRAN, Pascal, or C source programs respectively that exercise and demonstrate all of the *Numerical Recipes* programs. Each example program contains comments and is prefaced by a short description. Input and output data are supplied in many cases. The example books are designed to help readers incorporate procedures and subroutines and conduct simple validation tests. The example programs will not operate without the programs from the main book.

The programs contained in the main books and the example books are available in several machine-readable formats that will save users hours of tedious key-boarding. The IBM diskettes are $5\frac{1}{4}$-inch double-sided/double-density diskettes, which operate on DOS 2.0 or later on IBM PC compatible machines. The Macintosh disks are $3\frac{1}{2}$-inch single-sided disks for the Apple Macintosh.

Instructions

To obtain the books or the latest version of the disks, please order from your bookstore or complete the information on the order form found at the back of this book and mail it (or a photocopy) to Cambridge University Press in Port Chester, New York or Cambridge, England. Please note that there is a separate order form for each location. All orders must be prepaid. Prices are not guaranteed. Ordinary postage for shipping is paid by the publisher.

NB: Technical questions, corrections, and requests for information on mainframe and workstation licenses should be directed to Numerical Recipes Software, P.O. Box 243, Cambridge, MA 02238, U.S.A. Please do not write the publisher.

There are no cash refunds for diskettes. Only diskettes with manufacturing defects may be returned to the publisher for replacement.

ORDER FORM (Outside North America)

Order from your bookstore or mail to
Customer Services Department, Cambridge University Press, Edinburgh Building,
Shaftesbury Road, Cambridge CB2 2RU, U.K.

.... 38330-7 Numerical Recipes: The Art of Scientific Computing (FORTRAN
Version) £27.50

.... 30958-1 FORTRAN Diskette (IBM) £20.00

.... 35469-2 FORTRAN Diskette (Macintosh) £25.00

.... 34642-2 FORTRAN Diskette (Acorn) £20.00

.... 31330-9 FORTRAN Example Book £17.50

.... 30957-3 FORTRAN Example Diskette (IBM) £20.00

.... 35468-4 FORTRAN Example Diskette (Macintosh) £25.00

.... 34646-0 FORTRAN Example Diskette (Acorn) £20.00

.... 37516-9 Numerical Recipes in Pascal: The Art of Scientific Computing £27.50

.... 37532-0 Pascal Diskette Revised (IBM) £20.00

.... 38766-3 Pascal Diskette Revised (Macintosh) £25.00

.... 37675-0 Pascal Example Book Revised £17.50

.... 37533-9 Pascal Example Diskette Revised (IBM) £20.00

.... 38767-1 Pascal Example Diskette Revised (Macintosh) £25.00

.... 35465-X Numerical Recipes in C: The Art of Scientific Computing £27.50

.... 35466-8 C Diskette (IBM) £20.00

.... 37129-5 C Diskette (Macintosh) £25.00

.... 35746-2 C Example Book £17.50

.... 35467-6 C Example Diskette (IBM) £20.00

.... 37128-7 C Example Diskette (Macintosh) £25.00

Name .... (Block capitals please)

Address ....

....

....

Please accept my payment by cheque or money order in pounds sterling:
I enclose (circle one) a Cheque (made payable to Cambridge University Press)/UK Postal
Order/International Money Order/Bank Draft/Post Office Giro.

Please accept my payment by credit card:
Charge my (circle one) Barclaycard/VISA/Eurocard/Access/Mastercard/Bank Americard/
any other credit card bearing the Interbank symbol (please specify).

Card No. .... Expiry date: ....

Signature .... Date: ....

Address as registered by card company: ....

....

....

Prices of diskettes do not include V.A.T., which should be added to all U.K. purchases. 4-90

Order from your bookstore or mail to
Cambridge University Press, Order Department, 110 Midland Avenue, Port Chester, New York 10573

.... 38330-7 Numerical Recipes: The Art of Scientific Computing (FORTRAN Version) $47.50

.... 30958-1 FORTRAN Diskette (IBM) $32.50

.... 35469-2 FORTRAN Diskette (Macintosh) $32.50

.... 31330-9 FORTRAN Example Book $24.95

.... 30957-3 FORTRAN Example Diskette (IBM) $24.95

.... 35468-4 FORTRAN Example Diskette (Macintosh) $24.95

.... 37516-9 Numerical Recipes in Pascal: The Art of Scientific Computing $47.50

.... 37532-0 Pascal Diskette Revised (IBM) $32.50

.... 38766-3 Pascal Diskette Revised (Macintosh) $32.50

.... 37675-0 Pascal Example Book Revised $24.95

.... 37533-9 Pascal Example Diskette Revised (IBM) $24.95

.... 38767-1 Pascal Example Diskette Revised (Macintosh) $24.95

.... 35465-X Numerical Recipes in C: The Art of Scientific Computing $47.50

.... 35466-8 C Diskette (IBM) $32.50

.... 37129-5 C Diskette (Macintosh) $32.50

.... 35746-2 C Example Book $24.95

.... 35467-6 C Example Diskette (IBM) $24.95

.... 37128-7 C Example Diskette (Macintosh) $24.95

Please indicate method of payment: check ...., Mastercard ...., or Visa ....

Name ....

Address ....

....

....

Card No. .... Expiration date ....

Signature ....

.... Please indicate the total number of items ordered,

.... total price,

.... tax, if applicable (NY and CA residents)

.... total enclosed

4-90